A Guide to Understanding Healing Plants

Volume I

A Guide to Understanding Healing Plants

Volume I

Jochen Bockemühl

Translated by Harold Jurgens
Translation fully revised by David Heaf

MERCURY PRESS

Originally published under the title *Ein Leitfaden zur Heilpflanzenerkenntnis*, 1996,
by the Science Section of the School of Spiritual Science, Goetheanum, Dornach, Switzerland

Layout: Georg Iliev, Sofia

ISBN 978-1-957569-22-2

MERCURY PRESS
an imprint of SteinerBooks
834 Main Street, PO Box 358
Spencertown, New York 12165
www.steinerbooks.org

Printed in the United States of America

Contents

Contents

Foreword

Every young doctor who is interested in anthroposophy has the urgent question: "How can I understand the healing effect of natural substances, especially plants?" For many this question then becomes submerged in the immediate demands of clinical practice. In recent years, however, awareness of this situation has given rise to various initiatives to allow for a more reliable way of approaching healing plants and the preparation of remedies. From the outset, this was a matter of concern for Jochen Bockemühl in his work at the Science Section of the Goetheanum. He has researched healing plants for many years and worked with doctors, pharmacists and nutritionists on Rudolf Steiner's medical courses, including the book *Fundamentals of Therapy* which Rudolf Steiner co-authored with Ita Wegman. As a result, he is well acquainted with cognitional problems in medicine.

For decades, Jochen Bockemühl and his colleagues have developed an approach which involves placing themselves in the natural world with all their senses, of moving with it, and of observing and schooling their feelings and thoughts through it. Thus they have become able to describe the action of the various ethers and make them understandable conceptually for those familiarizing themselves with them (summarized in *Towards a Phenomenology of the Etheric World*, Anthroposophic Press, 1985). The approach in the present book is a continuation of Goethean methods, starting from Chapter 1 of *Fundamentals of Therapy*. Strengthening thinking is the first step in the training described there. Through it the relation of thinking to perception changes, at first from an identification of objects, into a 'capacity for perception'. In this way objective things gradually become transparent for the soul and spirit, and so does the inner nature of substances and of human organs. This method makes it possible to reach results in solo or group exercises more easily than through the written word. So it is better, with the help of this book, to seek the sources that made these indications possible and not ready-made guidelines or instructions.

Initiatives out of such approaches are crucial for the progress of anthroposophical medicine. So too are investigations of individual substances, plants or remedies that cannot be done by everyone but only by small, scattered groups on behalf of the whole medical movement.

Many thanks to Jochen Bockemühl and his working group for their exemplary collaboration in this field.

Dr. Wolfgang Goebel, Herdecke Hospital, 1999

In speaking about the senses, our attitude toward the spirit should be one of waiting to see to what extent an indication of the spiritual results naturally from sensory observation. This spiritual aspect should be neither denied nor presupposed; we must wait and see if it shines in.

Rudolf Steiner, *Anthroposophie. Ein Fragment aus dem Jahre*, 1910 (GA 45), Dornach, 3rd edition 1980, p. 29; trans: *Anthroposophy (A Fragment)*, (Spring Valley: Anthroposophic Press, 1996) Ch. 2,
'The Human Being as a Sensory Organ' p. 86

Defining the Task

People have always tried to gain knowledge of healing substances and, in doing so, the way in which they have viewed the connection between man and nature has continually changed. When Rudolf Steiner founded an approach to medicine extended by anthroposophy in the first quarter of the 20th century, it came at a time when traditional views were beginning to waver because of developments in science. It was also becoming necessary to gain a new understanding of health and disease in a human sense, one that offsets a merely chemical-physiological view. Since then people's consciousness and their living conditions have changed dramatically.
As a foundation, we will take the preliminary work that Rudolf Steiner did and described from an anthroposophical viewpoint and develop it. As in every science, it is necessary to get an overview of the particular field from the experts. Becoming familiar with this field is a condition for being able to take anthroposophical ideas as working hypotheses and then combining them with scientific facts. However, the anthroposophical approach also requires us to tread new paths with new methods and to be receptive to percepts other than those we are used to, i.e. we set out on our own path of cognition.
After a suitable preparation, the path taken here leads from outer observation to inner perception and so to the human being.
Therefore one of the aims of this guide is methodically to bring about percepts of the connection between people and natural substances and, in this way, awaken a clearer awareness of how feeling judgments arise and what they are directed at. To begin with this requires us to extend our ability to perceive our own thinking activity and to reflect on the movements it makes as well as on the assumptions that are made in every act of comprehension. This method brings life processes into our consciousness.

The aim of this process is to arrive at a perception of a substance's particular direction of activity that is permeated by our own experiences. While we follow a stimulus to go in certain directions and to reflect on the quality of the differentiations perceived in doing so, in science we necessarily have to transform the capacity of self-control that is exercised in the judgment of facts into an inner observer who accompanies us on the path in such a way that it remains easily comprehensible.
Such an approach can help a doctor to find the direction of activity of a substance that can be related to his perception of a particular patient. The same applies to pharmacists who have to turn natural substances into a form suitable for administration in such a way that the essence of a substance is active in the purified form.
Thus it is a matter of gradually arriving at the rationale that Rudolf Steiner promoted in a pharmacy extended by anthroposophy. Of course, this will have to be complemented by the work of others.

Premises and Aims

In working on the question of the extension of science through anthroposophy, I became aware of the risk of often having an idea of the path of anthroposophical schooling that is much too abstract, and of therefore thinking that the experiences described by Rudolf Steiner can only be taken as revelations. The statement from Rudolf Steiner's *How to Know Higher Worlds*, shown alongside, became a guide for me.
Steiner points out in his epistemological writings how an awareness of reality arises through the union of the two experiences of perception and thinking. The anthroposophical schooling path enables us to develop organs of cognition at other levels of reality. In *Anthroposophy*, a manuscript written in 1910, an attempt is made to describe the connection of these levels, proceeding from sense observation. Perception is extended by inner experiences that are connected with self-experience. Steiner wanted to

Above all, we must eradicate the belief that only bizarre and mysterious practices lead to higher knowledge. We should be clear that development begins with the feelings and thoughts we live with all the time, but that these feelings and thoughts must be given a new, unaccustomed direction. . . . In the final analysis, everything is based on the simple fact that, though we carry body, soul and spirit about with us, we are conscious only of the body.

Rudolf Steiner, *Wie erlangt man Erkenntnisse der höheren Welten?* (GA 10), Dornach, 22nd edition 1975, p. 59; trans: *How To Know Higher Worlds* (Spring Valley: Anthroposophic Press, 1994) Ch. 2 , 'Controlling our Thoughts and Feelings' p. 55

Introduction

In a lecture Rudolf Steiner gave on 22 July 1921, he indicated the method by which he arrived at his theory of the senses:
'We should acquaint ourselves with the thought that the theory of the senses should not be treated in such a way that we only describe senses according to the coarser sense organs, but according to an analysis of the field of experience. That is, previous investigators had only assumed the existence of a sense where they could easily find a sense organ; for these people an eye and an ear were the prototypes of senses.' Steiner points out – and physiology of the senses has confirmed it in the course of the 20th century – that there are further fields of sensorial experience: a sense of movement; a sense of warmth, etc. People have now discovered the physical sense organs of these senses, likewise with the sense of balance. But we should always be aware that direct experience, for instance the experience of warmth, occasions us to look for a physical organ for the sense of warmth. We know a sense through direct inner experience, and from the experience of the sense we pick out a particular organ for this sense. But when we proceed thus we get into certain difficulties that do not seem to arise if senses are defined by proceeding from the sense organs. For we now have to school our inner perception, inner observation, to learn to distinguish the individual sensory fields from all the others.
Two elements must come together if we want to talk about a sensory field. One is alert sensorial activity in the soul, i.e. as hearing and listening, seeing, smelling or tasting. The other is that this soul activity comes to meet each particular aspect of the world that makes an impression on us, be it sound, color, etc.'

From Christoph Lindenberg's afterword to his compilation of lectures about sense theory by Rudolf Steiner, *Rudolf Steiner – Themen aus dem Gesamtwerk 3 – Zur Sinneslehre*, Freies Geistesleben, Stuttgart, 1990 3rd edition, p. 135

disclose man's whole organization from within like this. The senses, the life processes, the processes in the inner human being and the 'I' experience are differentiated through an 'analysis of the field of experience', thereby enabling knowledge of the inner nature of human organs. This was intended to create the right basis to enable anthropology to discover the physical organs for the concept sense, for instance. The 1910 manuscript remained a fragment.
This approach was later continued in a changed form, with the foundation of a medicine exteded by anthroposophy, as the starting point for a new science.
In working with the book *Fundamentals of Therapy*, by Rudolf Steiner and Ita Wegman, it became increasingly clear to me how important is its first chapter, which deals with steps in the cognitional path from current scientific views to imaginative, inspirational and intuitive cognition with respect to an understanding of substances that have a healing effect. When we read the book's later chapters we can easily lose sight of the importance that this first chapter has for a real understanding of them. And in fact, Rudolf Steiner only added it later, with the obvious intention of clarifying the conditions for an understanding of the following text.
This guide starts with a clarification of the path that is described in that chapter. It gives examples of exercises that make personal experience possible, making thinking capable of seeing, and opening our eyes to perception at various levels. This creates a foundation for work with various cognitional approaches as well as with individual plants and substances.
Chapters I to VII of the book are mainly about our latest *methodology* for an understanding of healing plants. In Chapter VIII there is a description of a *practical* application of these methods to two examples. These plant studies, each case from a particular viewpoint, are continued in subsequent volumes of this guide.

I would be happy if the reader looked upon this guide as working material and if an exchange of experiences about it could take place. I would really welcome this.

Origins and Colleagues

Looking for an understanding of the inner connection between people and natural phenomena, which is the basis of this work, has been important in my life since my youth. My professional background for this guide is a study of biology and many years of engagement with this theme, which I took up in 1956 at the Goetheanum Research Institute. Corresponding investigations were also partly carried out as training projects by people who were studying science at the Institute. Work with biologists, mathematicians, farmers, pharmacists and doctors had already taken place before that. In addition, there were annual practical courses for pharmacists and doctors at the Institute from 1968, as well as workshops with scientific co-workers from Weleda Schwäbisch-Gmünd and Arlesheim.

Expression of Thanks

This guide for understanding healing plants arose through collaboration with a group of co-workers in the Goetheanum Research Institute: Torsten Arncken, Ulrike Ortin and others; from Weleda, Helga Eiben, Willi Hauert, Jürg Himmelsbach, Manfred Kohlhase, Monika Mennet-von Eiff, Albert Schmidli, Martin Schüpbach and others, and with the doctors and pharmacists Tillmann Feuchtinger, Wolfgang Goebel, Gottfried Klemp, Johanna Lobeck, Gerhard Schaller, Konrad Urech and others. Weleda AG also supported the work financially.

Mathias Buess, with the help of Silvia Briner and others, prepared the illustrations for printing and created a large number of them himself.
I would like to express my warm thanks to all of them here.

Finally, I should like to thank Gerald Karnow for providing the impetus for producing an English translation; Harold Jurgens and David Heaf for carrying it out; Pat Cheney for her meticulous copy-editing of the English text and Georg Iliev for design and layout.

How to Read this Book

The main text is generally in the two inner columns, whereas complementary extracts, references, notes, and illustrations are mostly in the two outer columns.

When we look at old and new popular literature on healing plants, we can see that during the course of the 20th century, there was a big change in the way people looked at substances with a healing action. This change is one that is important for modern consciousness. It is clear that until the 1920s people wanted to describe plants and their properties in experienceable pictures and to pass on traditional knowledge. This approach has increasingly been replaced by the kind of modern thinking about substances that is obtained in physics, chemistry and empirical physiology and which is based on models. We will give two examples from the literature, in which a plant is treated in a different way in each case, to show what the transition from one way of looking at things to the other consists of.

Handed Down Experiences and Attempts to arrive at an Experienceable Picture

In the book *Kräutersegen (Herb Blessing)* by E. M. Zimmerer (first edition, 1896), the presentation of yarrow begins with a description of the way in which it appears with other white flowers in summer meadows. Appealing to our feelings, the text goes into yarrow's many common names and what has been handed down in sagas and legends. For instance, Achilles knew that he could heal Telephos' wound with yarrow. A list of its uses follows. At the end of the introduction we read: 'Yarrow's healing power is largely based on its almost closed leaves and its fairly strongly smelling flowers whose components mainly comprise tannin, essential oils, salts and bitter extractable substances. It belongs to the bitter, tonic, strengthening, cramp loosening, anti-flatulent, and stimulating remedies with respect to its effects. Its use is seldom without success'.
This description gives the readers, who presumably already know the plant to some extent from everyday experience, a very varied and attractive picture of yarrow, and the detailed description of its uses which follows shows the large number of ways in

which people once used the plant to heal directly perceived complaints that were often hard to establish more exactly. Indications about the gathering time and preparation follow. We hear something about traditional 'experiential medicine' and get a general idea of how Paracelsus and others before him lived with imaginative pictures that were still united with the corresponding clinical pictures. Of course, we only get a slight inkling of such inner pictures by reading *Kräutersegen*. They are reflected more in the manner of their presentation.

Illustrations in books from around the beginning of the 20th century were mostly very exact and naturalistic, whereas in earlier times (for example in De Simplici Medicina, published at the end of the 14th century) plants were illustrated more symbolically. Obviously figures did not have the same significance then as in the age of photography.

Mixed in with the descriptions in Zimmerer's book are concepts from a new way of thinking which is mainly directed to 'components' that are distinguished first of all through direct perception, through the way substances are obtained from plants or through their special usage. By 'bitter extractable substances' we can picture straight away something that can be extracted from plants and that tastes bitter. 'Ethereal oils' are slightly volatile and can be recognized by their strong odor. They can be obtained through distillation. 'Salts' are left behind in the ashes and can be further differentiated through taste. Extracted 'tannins' have an astringent taste and are used to tan leather. The name of the component substance expresses what we can produce with it and make perceptible.

Symbolic pictures of yarrow and chamomile from a medieval herb book (*De Simplici Medicina*, end of 14th century, Basel University Library).

The remedy from yarrow has an essential oil in it (ca. 0.2–0.5%) that contains pinene, camphor, bornyl acetate, sabinene, artemisia ketone, cineole and other terpenes. The bitter tasting proazulene, matricin, is only found in the essential oil of tetraploidic varieties of the one mentioned above. Other ingredients are flavonoids and tannins.

(Hans P. Dörfler and Gerhard Roselt, *Heilpflanzen*, Frankfurt, 5th edition, 1990)

Modern Understanding of Substances and Thinking in Terms of Models

The current way of thinking comes very much to the fore in our second example, the booklet *Heilpflanzen (Healing Plants)* by Hans P. Dörfler and Gerhard Roselt. Here the presentation is strictly systematized, whereas in *Kräutersegen* it resulted more from the author's personal relation to plants. Now the 'findings' are presented in a systematic form. We find a brief description of the properties, occurrence, and range of yarrow as well as what to gather, when and how. Everything that was just copied from old books is left out, but so is everything that appeals to our feelings. A list of the substances it contains is now the main thing.
Nowadays direct perception is becoming less and less important in characterizing such substances when people speak of them. They are described in terms of the way they react chemically. Structural formulae are ultimately a symbol for this. Ever more substances can be isolated and differentiated through ever more complicated methods. The same substances can often be obtained from different plant species, for example, matricin from yarrow and chamomile, cumene from chamomile and cumin, or they can be synthesized from already isolated substances. The formulae no longer tell us anything about the plant from which the substances come and vivid descriptions of plants can no longer be connected with them. Thus, as a rule, yarrow the plant no longer stands in the foreground but rather the chemistry of its isolated substances.
This development is accelerating. Ten years after *Heilpflanzen* was published, the reality of the plant had been almost completely replaced by formulae and active substances in the consciousness of people who are scientifically oriented.
A connection with human beings is established by animal experiments and clinical tests. Thus, physiological processes become measurable with the help of experiments and thinking in terms of

models, and are explained by mechanisms. Life functions and consciousness states can be influenced and controlled through such knowledge and by mastering special processes in the organism that are considered to be mechanisms.
But this knowledge does not enable us to arrive at a full understanding of healthy or sick people and as a result our concept of disease has become an immense problem.

The Path to the Art of Healing through a Methodical Extension of Direct Experience

Rudolf Steiner anticipated and affirmed the development indicated by our comparison of the two books on healing plants, i.e. from a more pictorial experience of the healing connection between natural substances and human beings to modern model-based, technical thinking that has a merely empirical connection with human organ functions. In view of this distancing tendency, he considered it necessary to develop a complementary knowledge that proceeds from a direct experience of human beings and seeks out a personally convincing concept of disease as well as the healing relationship between a person's nature and the substances around him. That is why he worked with doctors, in particular with Ita Wegman, on the foundation of an extended medicine. The book *Fundamentals of Therapy* (Spring Valley: Mercury Press, 1999) contains a brief description of the cognitional path necessary for this and a new concept of disease. In addition, illustrations that result from such modes of cognition are formulated in this book and in many lectures. Work with them eventually led to the development of anthroposophical medicine in current practice.
But even this is undergoing change. Therefore it is necessary to develop a new approach to an understanding of the healing connection between man and nature out of present day needs for knowledge. We should understand the following chapters in this sense.

This small book presents new approaches in medical knowledge and skills. A proper judgment of its contents will only be possible for those who are prepared to consider the points of view that were dominant when the medical views discussed in these pages evolved. It is not a matter of being in opposition to the school of medicine that is working with the accepted scientific methods of the present time. We fully acknowledge its principles. And in our view, the approach we present should only be used by those who are fully able and entitled to practise medicine according to those principles. We do, however, add further insights to such knowledge of the human being as is now available through accepted scientific methods. ...we therefore feel compelled to work for an extension of clinical medicine, based on these wider insights into the nature of the world and the human being.

Rudolf Steiner and Ita Wegman, *Grundlegendes für eine Erweiterung der Heilkunst nach geisteswissenschaftlichen Erkenntnissen* (GA 27), Dornach, 5th edition 1977, p. 7; trans: *Fundamentals of Therapy* (London: Rudolf Steiner Press, 1996) Ch. 1, p. 1

Deadly nightshade *Atropa belladonna* L.

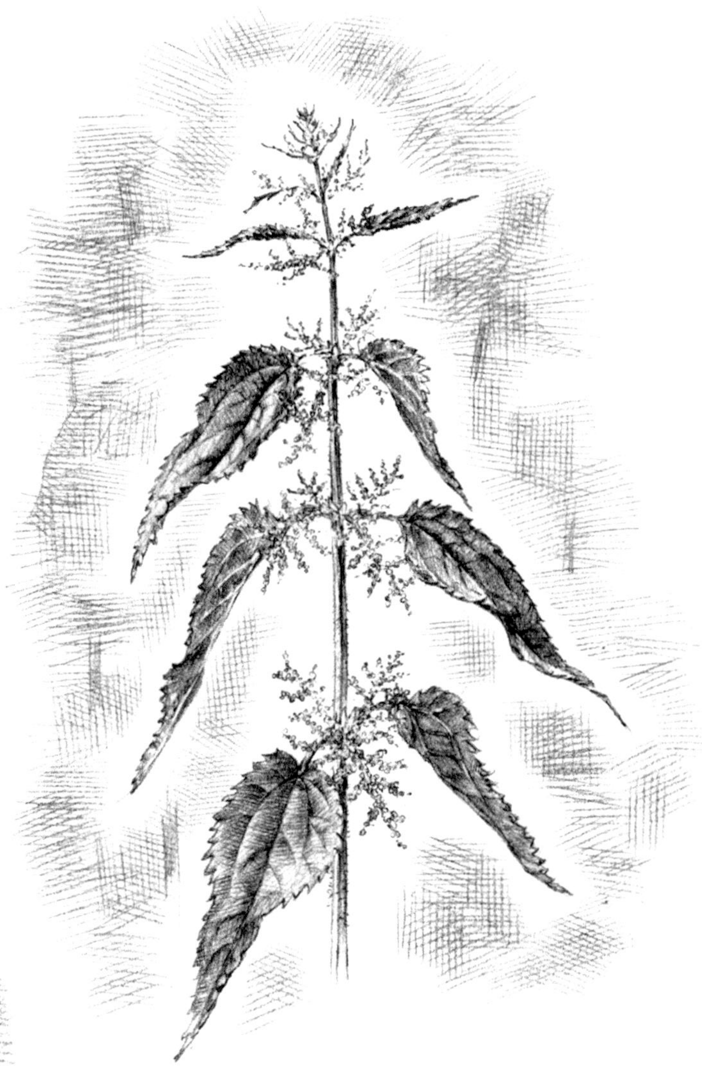

Nettle *Urtica dioica* L.

II Unconscious Connections of Humans & Animals with Substances

How out of touch we are with plants! We become aware of this if, when away from our kitchens and shops, we ever ask ourselves whether a particular fruit, root or leaf could be used for eating or for medicinal purposes. The berries of belladonna are attractively glossy and even taste pleasantly sweet. They do not look very poisonous. Stinging nettle, however, has a directly perceptible effect: we feel a stinging sensation as soon as the plant is touched. But this does not tell us anything about its healing effect. When we touch figwort it emits an unpleasant smell that stops us from eating it. But could we use it as a remedy? (People used to take it for scrofulous ailments.)

We feel a similar uncertainty about the healing effect or edibility of most substances around us so long as we only confront them perceptively and judge them as external things. On the other hand, there are natural, active connections between people or animals with substances in their surroundings and we assume that such connections exist even though we are usually not aware of them. They form a wise foundation for our whole life.

Common figwort *Scrophularia nodosa* L.

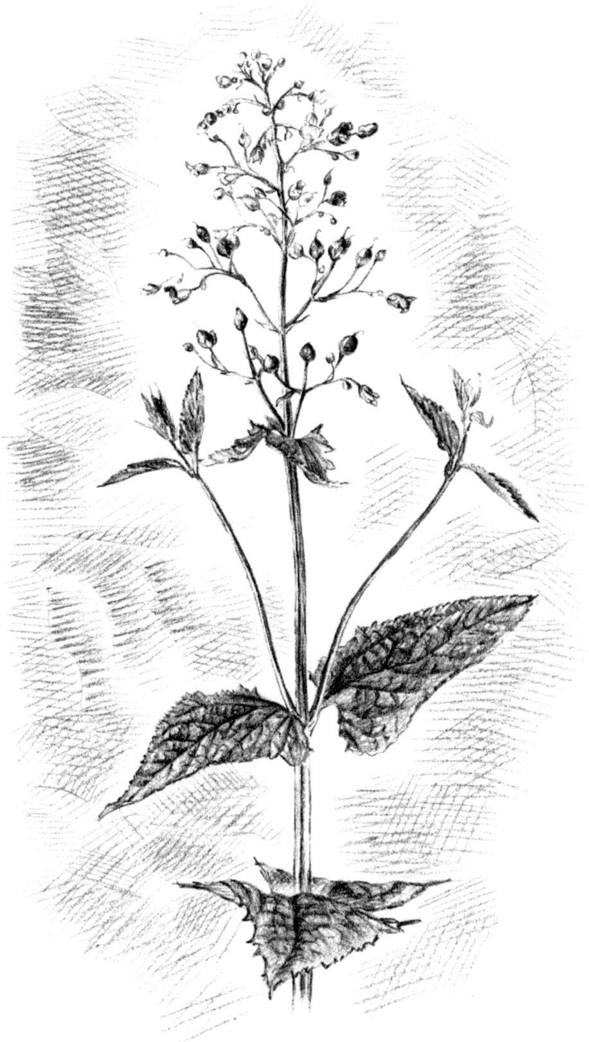

Instinctive Relationships of Animals with Substances in their Environment

We are not surprised that animals eat what they like and avoid things that interfere with their feeling of well-being. We assume that there is a direct connection between them and certain plants. A cow eats the grass around poisonous buttercups without trying them. Squirrels nibble nuts, but not grass. It is obvious to us that nuts belong to squirrels as grass belongs to cows.

If we observe how a cow grazes with her muzzle close to the ground, touching and smelling plants and grass, tearing them off with her tongue and eating them while leaving the buttercups alone, then we can really see that she lives entirely in smelling and tasting and behaves accordingly. She does not 'know' things as we do. Instead, her sense impression passes directly over into behaviour. For us, the cow's behaviour has become an expression of her that we see and experience.

Do animals learn to look for or avoid certain plants, or are they born with this behaviour? The answer is clear with a calf – it goes straight for the udder, and does not at first try to eat grass or catch mice. The inner connection is inborn. A capacity to digest food is also connected with this. When the digestion changes, so does the behaviour, and other food is eaten. On the other hand, habits can work back upon receptivity. For example, a very fibrous fodder educates the digestive capacity.

Occasionally something unexpected happens: a cat that at other times eats mice suddenly consumes grass. Apparently it has to do this to stay healthy when it is in a particular condition. Here too are clear soul connections through which an organism is connected with substances in its environment. The surroundings obviously belong to an animal's organization as much as the particular form and function of its outer and inner organs.

Although a cow can avoid buttercups and can change its behaviour in certain life situations, it can never confront its actions consciously.

Even if a cow includes her calf in her soul surroundings and 'takes care' of it, she obviously does not develop any experience of the reality of the calf she is faced with. She acts as if she were only living in pictures that arise and direct her actions, and that are only stimulated by what she senses. We cannot speak of self-consciousness here, only of a dreamy one in which perception and inner experience form a whole from which action and behavior flows. That is what we usually call instinctive behavior.

Instinctive Relationships of Human Beings with Plant and other Substances

People also have perceptions that tend to happen dreamily and that pass over into actions without being mediated by waking consciousness. A good appetite is considered an important basis for a healthy life. When we eat, sympathies and antipathies for particular substances arise that begin with seeing and smelling and continue with touching and tasting in the mouth and gastrointestinal region. Our organism's life is based on a healthy functioning of these processes. To strengthen ourselves it may be necessary to overcome the antipathies that arise when we look at a particular food, but not always; if we have a very strong aversion for mushrooms etc., forcing ourselves to eat them can cause vomiting. The organism protects itself against something that is intolerable for it. It 'tastes' it in its lower regions, which ordinarily remain unconscious. It is often possible to observe our organism reacting to perceptions with specific physiological processes. We can access this most easily in after-images; for instance if we look at a red surface and then at a white one, we see a greenish image that moves with the eyes.

To recognize this as an objective process we have to concentrate on it and avoid arbitrary distractions. For instance if we see someone preparing lemon juice something in our mouth constricts. Our sensations are quite different from when we see chocolates that give a foretaste of sweetness, or when we look at salty pretzels. Each perception connects with particular sensations that pass over into reactions of the organism. Conversely a more or less directed striving towards particular percepts can appear as a 'need' or in pathological cases can become an addiction.

Changeability of Instinctive Relationships

The desire to eat something sweet like honey, to drink water or to eat bread is based on quite different sensations. That is why we reach for honey when we are full and perhaps a little tired, or for a glass of water if we have talked a lot in dry air, or we cut a slice of bread if we come home hungry.

Although we usually avoid bitter things, we occasionally feel a desire for wormwood tea. This is obviously connected with discomfort in the organism. The need itself is usually much clearer than the knowledge of which organ it is coming from.

These things depend a lot on the organism's condition. Sensations change continuously during drinking and eating. This gives instinctive action a different direction depending on the circumstances. If I have a desire for something sweet and I eat a spoonful of honey, it will significantly reduce the desire to take more. If I eat another spoonful anyway, it may produce an aversion.

One can also look at the relation between foods and spices in this way. What is meant here as the instinctive process in the organism that connects the soul with substances is always specific in each case. Rudolf Steiner called such things life processes.

Life Processes Regulate Relationships with Substances

In *Anthroposophy, A Fragment* (GA 45) Rudolf Steiner says that life processes are a bridge between sense percepts and our own experiences. Connections like judgments affecting the soul arise from perception via the sensation of life processes before waking self-consciousness comes in and recognizes a percept as something external. This recognition distinguishes people from animals. Inner and outer confront each other as separate things only through self-conscious cognition, whereas the life processes that bridge the two usually remain unconscious. This confrontation facilitates the clear recognition 'from outside', but it also generates the uncertainty already described in recognizing the real, active connections. The ability to distance ourself also tends to change our natural instincts. Thus a food that one person might eat with pleasure could arouse disgust in someone else who thinks of dirt and pathogens when he looks at it.

Rudolf Steiner, *Anthroposophie. Ein Fragment aus dem Jahre*, 1910 (GA 45), Dornach, 3rd edition 1980; trans: *Anthroposophy (A Fragment)*, (Spring Valley: Anthroposophic Press, 1996), (see also Chapter VII, passim)

To summarize, there are active connections between live creatures and substances that are mediated by 'pictures' of the environment in which these live creatures are embedded. This phenomenon is particularly developed in wild animals. Such connections with substances are also active in human life, but they are often changed by the thinking consciousness in such a way that they are no longer directly accessible to medical, pharmaceutical knowledge.

It is one of the aims of this guide to stimulate new perceptions of these connections, and in this way awaken a clearer consciousness of how an instinctive feeling judgment arises and what it aims at. We become aware of life processes, the understanding of which is necessary for an extension of pharmaceutical knowledge, through the exercises suggested in what follows.

III Developing the Perceptual Capacity of Thinking

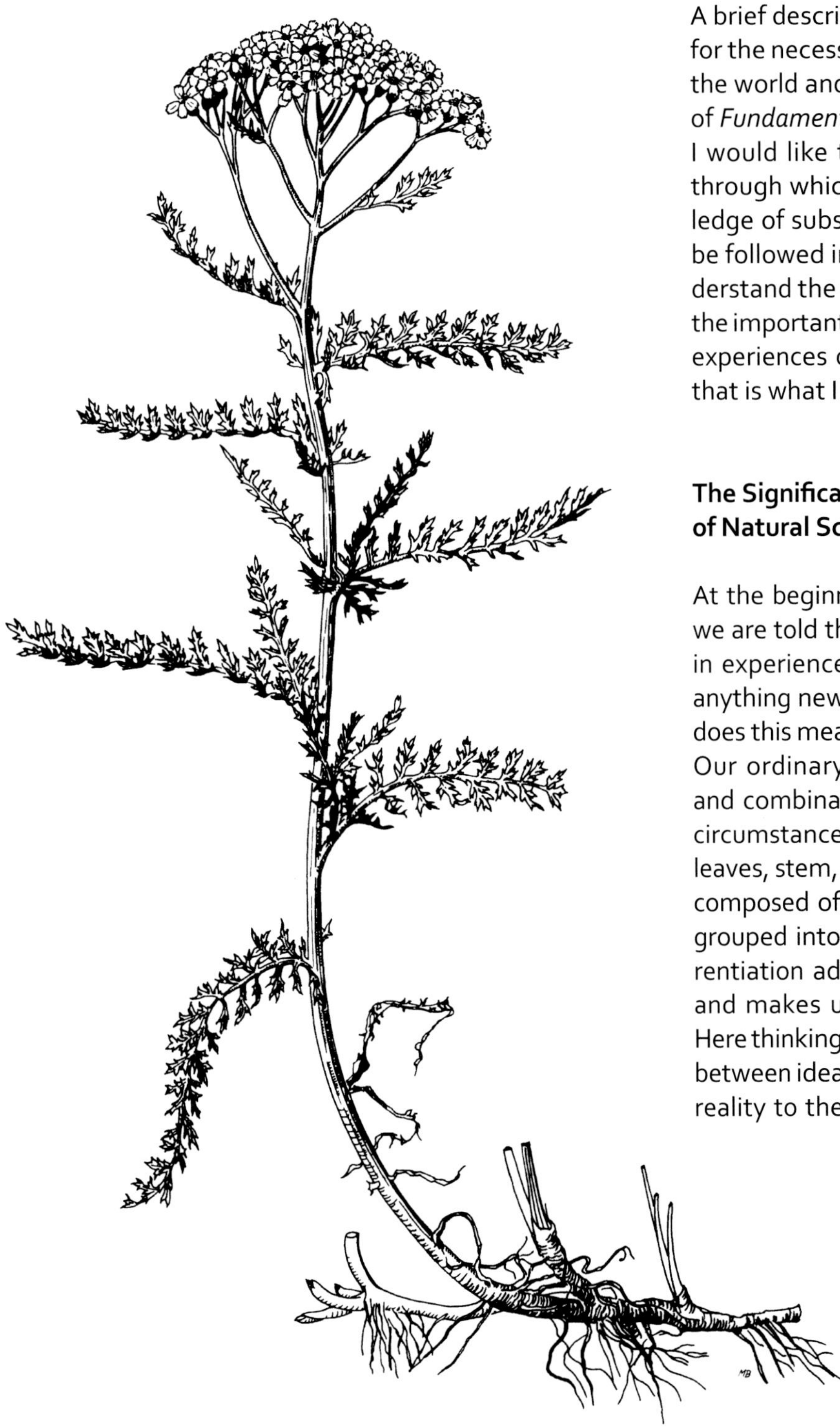

A brief description of the steps that must be taken for the necessary extension of an understanding of the world and mankind is given in the first chapter of *Fundamentals of Therapy*. Following these steps I would like to describe a sequence of exercises through which the tools for an appropriate knowledge of substances can be created. This path can be followed initially only in thought in order to understand the reflections that follow. But in the end the important thing is to arrive at the corresponding experiences one's self through the exercises. And that is what I would like to encourage you to do!

The Significance and Limitations of the Methods of Natural Science

At the beginning of the chapter mentioned above we are told that the ordinary thinking that is active in experience of the physical world does not add anything new to what is given by the senses. What does this mean?
Our ordinary thinking which makes separations and combinations brings consciousness into outer circumstances; for instance yarrow is divided into leaves, stem, flowers and so on. Flowers are seen as composed of calyx and corolla, several flowers are grouped into inflorescences, etc. Every such differentiation adds new concepts of the same quality and makes us aware of more outer connections. Here thinking mainly consists of making connections between ideas. From *this* viewpoint, it adds no new reality to the reality of the sense world, as it only

wishes to express the differences and arrangements that are formed there. This identification with the outer world is the step in consciousness through which we gain clarity in science.

In this we are not really conscious of either our own activity in perception and thinking or of the contents of perception. Perceptions are made in a brief moment. It is true that the 'discovery' lets something light up in thinking, but usually conceptual connections arise immediately that 'overshadow' the perception again. The thoughtfulness needed in order to enter a pure experience of a percept is usually lacking, and so a feeling of dissatisfaction and of not being able to get into it remains.

Therefore cognition remains shadowy compared with the perception of qualities or effects, because the latter live more in our feelings and only become properly conscious in the form of finished ideas. Hence an extension of knowledge of the world and man through anthroposophy's path of cognition must begin with a further *development of our thinking capacity* in the direction of these deeper layers of consciousness.

What is really meant by yarrow, leaf, flower and so on? I usually use these concepts without becoming aware of how I form them. Each concept can be used in very different ways. I say that it has several meanings. If I want to grasp and experience its full content, I have to stay awake and mobile in my thinking activity. Then I can discover that it only acquires its specific meaning in *each* context and yet it nevertheless remains a whole. To exercise this mobility I can look for the connections with many phenomena I find in a concept. In this way, by making the activity of thinking conscious and by experiencing its creative element, we begin to extend our knowledge.

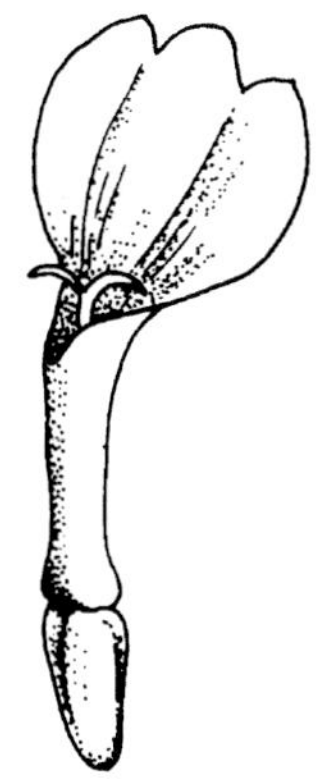

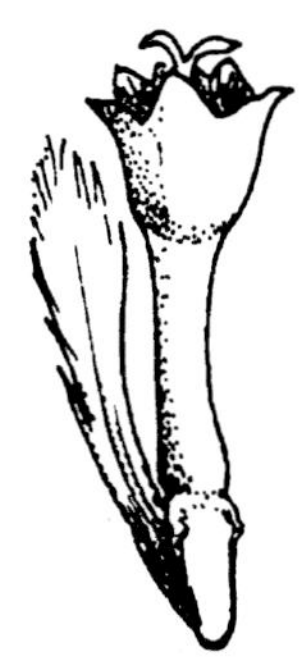

In that we understand the facts around us when we think about them, we arrive at certain boundary concepts such as atom, matter and so on . . . These boundary concepts, boundary ideas, which come to us especially when we follow the branches of science that underlie technology, stand there like pillars, so to speak. And if we want to remain with conventional science we will definitely remain standing in front of these delimiting pillars.

But for the spiritual investigator of the kind I mean here, the real work begins only at these boundary pillars. For the spiritual investigator it is a matter of entering into a certain inner battle, into an inner struggle of life with these concepts, more or less with all the delimiting concepts of natural science, in what I call – please do not be put off by this, it is a technical term just like the others – in what I call meditation. And this inner battle does not remain unproductive for him . . .

. . . What I am referring to here is not brooding introspection but rather that we consult everything within our soul for what led to these concepts, that we try to live with these concepts meditatively.

What does that really mean? It means establishing an inner discipline within and being able to look at what we have in our soul when we arrive at such a boundary concept, just as we otherwise look at outer objects. I could suggest a lot more of them besides the ones I have just mentioned.

Rudolf Steiner, *Geisteswissenschaft, Naturwissenschaft, Technik* (GA 107), Dornach, 1991, pp. 25–6 (See Beiträge zur Rudolf Steiner Gesamtausgabe, no. 107, lecture 17th June 1920)

Strengthening Thinking

A thought is only really present in our consciousness for as long as we actually produce it. Therefore we are usually only dealing with representations or 'shadows' of thoughts. We deal with these in a way similar to the way in which we deal with objects, i.e. as if we could interchange them without altering them. Our ordinary relation to substance is also based on this.

In contrast to this, the strengthening of thinking is done by directing consciousness at its own content. For instance I see a flower and identify it as yarrow, but then I do everything I can to let the content of 'yarrow' become awake in my consciousness or to keep it in motion. Thinking is usually only directed outward at a thing, but here my inner eye is also directed at my own thinking activity. I direct my attention not just to the thing that is foreign to me, but at the same time to what is produced within me to illuminate it; something I am only half-aware of in my thought life.

To strengthen our thinking and to arrive at an extended ability to perceive, we direct our attention to what we connect with a particular concept in various situations, and to what induces us to form it in a particular situation. If this conceptual connection is thought and experienced repeatedly, our thinking process intensifies. This leads to boundary experiences that form a transition to new inner experiences. Here we will call such exercises, in Rudolf Steiner's sense (1920), meditations in thought. In what follows suggestions for three mutually complementary exercises are given:

– 'picture' as content of meditation,
– a projective geometry exercise,
– the sensible-supersensible plant.

III Developing the Perceptual Capacity of Thinking

'Picture' as Content of Meditation

We can also let 'picture' become a 'boundary concept'. What do we mean when we speak of a 'picture' or of 'pictorial'? We use the concept 'picture' in quite different ways without usually becoming aware of it. *A picture on the wall* can be an object. We can *express something in a picture*, or be *in the picture* about something. Because we only have the one meaning in mind, we do not usually realize that each statement also contains a particular point of view that belongs to the inner nature of the expressed fact. So we take the concept 'picture', focus on various concrete situations in which we want to speak of picture or pictorial, and then formulate a statement in each case that corresponds to this situation. Then we reflect on and describe our own relation to it, make ourselves aware of the change in viewpoint as we go from one situation to the other and try to experience it fully.

A possible series of statements will be given in the following (left), with the related steps in thinking (right):

I can say, "A picture is hanging on that wall. I think that it should hang a little higher".

I notice in this statement that I take for granted what a picture is, but I look upon it only as an object with which something can be done.

Now I say, "A picture is an object that we usually set apart from its surroundings with a frame".

I notice that my gaze is directed towards this region that is set apart. This region only has meaning if it is taken to be not merely a physically outlined object.

I formulate this as follows; "The picture should be seen as a whole. Its purely material aspects need to be complemented".

I ask myself why I see the picture as a whole.

Thus I arrive at the next statement; "The picture is not the thing itself. It points to something that the observer experiences inwardly".

I reflect that I could say the same thing about a photograph that directs my memory or imagination to a person, an object or a landscape. In this respect, my attention is led back to what is sensorial, objective. But when an artist paints a picture he is not just interested in making such a copy.

I decide, "Even if an object is copied in an artist's picture, it also directs my gaze to something else that is only accessible to experience".

Thus the object copied in a work of art becomes a picture; it does not just mean 'itself' – it is also an expression of something else.

I see and experience, "What is raised to the level of a picture by an artist is the result of the interaction between his soul activity and substance that is sense perceptible. It becomes an expression of this interaction".

Viewed from this angle, I realize that how a work of art appears is the essential thing. The 'how' speaks through the substance and the object.

What this means for me is that: "The way in which the picture is offered to an observer demands an inner activity that, while looking at the work of art, can become perception".

Thus with my experience I am standing at the transition between a thing situated outside and a purely inner experience, i.e. between seeing things and pictorial perceptive thinking.

The previous statement means that, "thinking activity acquires a pictorial character in my own experience".

This shows me that this pictorial character should not be confused with ideas. Ideas only serve as an intermediary state on the way to an inner perception that cannot be put into ideas.

Therefore, "perceptive thinking is an experience that emerges in my innermost being and there, as when I do projective geometry, testifies to a spiritual, outer world".

Thus, in this perceptive thinking, reality is experienced in my own picture-making activity that directs itself at something other and opens itself to it in such a way that the outer is also experienced as inner.
In this way I direct my attention to a region from which I ordinarily get my memories, for example, a memory of a sense percept. I can best observe how remembering takes place when I am finding it hard to remember something. In that case I should not try to create the memory picture directly.

Thus my thinking has to lead my consciousness in the right direction and to make itself open for what rises from within me.

So I tell myself that, "My thinking activity must become an organ of perception".

Whether I become aware of pictorial experiences that arise from within by themselves or whether I clothe such experiences in pictures myself in order to make them communicable, I can call both sorts imaginations. Imaginations – pictorial perceptions or pictorial insights – come from the same region as memories. But they differ considerably from ordinary memory pictures. With the latter it is mainly a matter of again raising into consciousness something that was perceived with the physical senses. A feeling that "I was there" is experienced in connection with my body, i.e. with earlier sense impressions – whereas conscious imagination as a cognitional capacity directs my attention to the agreement between a freely arising inner experience and my own activity in the production of a mental picture which corresponds to that experience.

Idea really means pictures (Gr. ειδος = that which is seen, a class or kind).

Something else results from this, "There are insights that arise without my help, that are not ordinary memories and that can only be retained if I transform them into mental pictures".

This leads me to the statement, "What we call an idea is found in the direction of inner perception too".

As I begin to identify myself with the nature of the other like this, I let it arise in me anew and I can produce new phenomenal forms out of it via the formative activity of my thinking.

If I look for something in my own experience that indicates what is meant here, I could say, "to begin with, idea is an inner experience of the spiritual that I can grasp in thinking. I can also see it as an inner light through which I experience an inner in something outside me".
In thinking about the concept 'picture', I have now come to a boundary where I notice that each statement is moved into the background by the next. And I feel affected existentially if on such a meditative path the concept begins to contradict itself in me, for instance when contrasting picture, as an object that can be imagined in a certain way, with idea as a 'pictureless picture', i.e. the capacity to freely produce mental images of a particular kind. Instead

If we then try to concentrate on such concepts with our whole soul to the exclusion of all other experiences, we will make an inner discovery. And this inner discovery is somewhat distressing. For it shows us that from a certain point of life, of inner life, our concepts become something that grow by themselves in our soul, that behave differently, after such inner meditative work than they behave if we only take them as the result of outer observation. Thus, just as we observe in children growing up how certain organs which first appear more undifferentiated, become differentiated, just as we perceive the way in which organs grow, so too we feel in such a meditative devotion to the results of scientific experience as if an inner growth of the soul is taking place. And then comes the distressing aspect, that we say to ourselves: it is not through speculation, it is not through philosophical speculation that we go beyond what is called the boundaries of natural science, but through direct experience, i.e. by transforming what we have gained through thinking into an inner experience of our perceiving.

Rudolf Steiner, *Geisteswissenschaft, Naturwissenschaft, Technik* (GA 107), Dornach, 1991, p. 25 (See Beiträge zur Rudolf Steiner Gesamtausgabe, no. 107, lecture 17th June 1920)

of it being brought to a close with a definition, on which I usually try to support myself, it is now open. I have completely connected myself with it inwardly and have carried out a kind of turning inside out with the direction of my gaze. Thus, the concept has become alive and an organ of perception for inner and outer things.

We can expand such a contemplative exercise on a given concept in any way we wish. It leads us to the secret of comprehension in general. It shows us that we can direct our consciousness in two directions, namely, towards the perception of a thing 'out there', and towards an experience that arises 'in us' as something familiar within our own being. The two experiences should be separated and we only place ourselves consciously in the respective reality by joining them together in full presence of mind – we could also say by inversion of one into the other.
Analytic thinking that is directed outward pushes itself 'between' things, connects them in thought, and in this way creates an awareness of spatial and temporal conditions. It thus gains logical clarity, but it has the tendency to lose the real content of perception that lies in a special experience.
In contrast, perceptive thinking that is turned inward merges into the content, into the unity of experience. It runs the risk of wandering into 'incomprehensibility' if the 'self' cannot stay in it.
It is not a matter of adhering to this polarity but rather of turning thinking's ability to perceive towards sense percepts, i.e. of developing perception in seeing.
Exercises with the 'picture' concept show me how to handle pictorial views and forms of expression, so that I do not drown in incomprehensibility. They help to develop a capacity through which an open mood of expectation arises as a bridge between *outer* and *inner*, and with which I wait for an experience of the *inner* in the *outer*.
In this way my inner observer arising in thinking has an overview at any one time of where I stand spiri-

tually in contemplation: when I speak of 'picture', am I thinking of something sensorial? Do I mean a particular mental image? An idea? Or do I want to draw attention to an inner experience that arises in my own activity at the moment when I work with a phenomenon?
The more we reflect on commonly used central concepts employed as so called supports – 'plant' for example – the more mobile and perceptive our thinking becomes.

Exercises in Projective Geometry

Strengthening of thinking comprises reflecting on and experiencing our own thinking activity concerning something that is at first strange to us.
Reflection on the concept 'picture' enables us to discover the ability to place ourselves in various connections with the concept in such a way that it begins to become an organ of perception.
In pictorial comprehension we try to place in a spiritual environment, something sensorially given from which the thing perceived acquires meaning. This initially shadowy spiritual environment of the picture 'in me' can be recognized and experienced as the real spiritual environment of the picture 'out there'. On the other hand, thinking can be experienced as something that is pictorial to the extent that it makes an outer connection with an inner experience through which the former becomes comprehensible. Both processes lead to convincing experiences in the I and therefore have the character of reality.
Learning how to move around in a purely spiritual element and to carry out the inversion process experienced in picture meditation is easiest with the help of projective geometry – starting from elements produced in our own imagination.
Let us practise on a triangle: reflecting on a triangle begins with an idea that is illustrated with chalk on a blackboard.
We create a sense percept as a mental support for something that cannot possibly be a sense percept

See for example: Olive Whicher, *Projective Geometry*, (London: Rudolf Steiner Press, 1971); Lawrence Edwards, *Projective Geometry*, (Edinburgh: Floris Books, 2003); Louis Locher-Ernst, *Space and Counterspace*, (Fair Oaks: Association of Waldorf Schools in North America, 2003)

but has to be thought. A line in chalk on the board is a model in this sense. Three lines and their crossing points always give a triangle.

Now keeping two of the corners of the imagined triangle fixed, let the third move in a certain direction. The further it moves the harder it is to imagine the triangle concretely. We can only grasp the movement with our thinking and accompany it with our experience. In our thoughts we can follow the movement out to infinity and experience the consequences of doing so. These consequences are unexpectedly contradictory. Trying to imagine them is impossible. For instance, the two sides whose point of intersection is infinitely distant must become parallel. So does a point of intersection still have any meaning here?

To follow the movement of a vertex of a triangle we need to look at another aspect of triangle transformation. In the first part of our exercise, moving the vertex in a straight line necessarily involved a rotation of the corresponding two sides around their fixed points, until the third point was out 'at infinity' and the sides were 'parallel'. We can continue these rotational movements uninterrupted even after it is no longer possible to imagine the vertex. Now if we imagine the lengths of the sides as unlimited in both directions we notice that the vertex that had moved to infinity becomes imaginable again. It comes back from the other side. It is also 'reversed'. Such an inversion process enables me to recognize as triangles shapes that previously seemed to be quite different from an outer viewpoint.

In following the movement of my own thinking I become aware of a geometrical principle that I experience as just as real as that which is otherwise only applied to objects. Thus this spiritual reality takes hold of me from within in a way that is similar to the outer world taking hold of me via the senses.

III Developing the Perceptual Capacity of Thinking

Conclusions from the two Preceding Exercises

In this kind of geometric exercise, we learn to begin with mental images in our thinking and to bring them into movement with the intention of consciously extracting and experiencing their inherent consequences, in order to construct more things, for example, all sorts of triangular shapes.
In this way we move into pure thinking and make percepts in a purely spiritual element, i.e. in the purely connective activity of our own thinking. New contents of perception arise that are the content of our own human nature and of the world.
The first exercise we performed (picture meditation) was essentially an inversion. In it the more important things were the direction of our gaze and an experience of the content. In the second exercise we used elements that we created ourselves in our imagination. The ideas we start with in geometry have no sensorial content. Point, straight line, circle, and so on stand before us entirely out of inwardly grasped thoughts. But they can be transferred to the sense world. Their connection with outer reality lies in the fact that we can illustrate the imagined relationships (on a blackboard, or with objects) or we can find them realized between things or as borders of things. We turn our body towards them in space and thereby determine our location. We make a connection between mathematical thought and physical reality by measuring, counting and weighing. Thus we ignore all directly perceived qualities. The only percepts that go directly into the procedure are those from the lower senses (senses of balance, movement and life) which we use mainly in our own body, and from touching that accompanies all sense percepts but stays at the surface of things in the body's experience of itself. In relation to objects that are immediately outside us but are unknowns, touching gives us our body's experience of reality, which is what our self-awareness usually uses to support itself.
When we 'feel our way' into sense phenomena

through the experiences we have via our limbs, we experience that 'form' is something real, which follows geometrical lawfulness. We unite with what we perceive out in space and deal with it.
So we are continuously taking outer operations into our thinking and applying thoughts from our 'operation' of mental images to outer actions.
These transitions correspond to the relation of point to circumference. We should realize that sense percepts can only be made at a point in space and time, i.e. only here and now. Sense observation gives us experiences of 'next to each other' and 'one after another'. They always have mutual spatial and temporal connections on which we and our body are dependent. We can think these connections but with our thinking, we are at the spiritual circumference of sensorial things, i.e. outside space and time. Of course we can only think 'now'. But there does not need to be anything out there in space for this. Looked at this way it is immaterial when and where we think something or where the object that the thought is connected with is located. What we experience in this way is completely spiritual.
In science we create transitional pictures, i.e. models, to make a connection with an object. They illustrate spatial-temporal principles without standing in the reality of space themselves. Just as these models (pictures) remain outside sensory qualities and only make spatial and temporal connections between them, so our experiences in mathematical-geometric or logical trains of thought make spiritually real things inwardly perceptible. We may also find that this power of perception can be considerably strengthened and expanded through long-term regular exercises.
Of course we are still dealing here with what is purely formal and spiritual. However, when it is used as an instrument (as shown with projective geometry) for converting mental images based on experiences from our limbs into pure thought permeated by outer experience, then such *thought movements* can become *guidelines*.

III Developing the Perceptual Capacity of Thinking

The next step in these guidelines is to carry out the inversion of the outer experiences themselves into inner ones in a way that is different from the one we became familiar with in the preceding exercise. This is also done with perceptual qualities in a way that is as exact as in mathematics. Such a step is also necessary for reaching an understanding of the inner nature of substances. This can be further prepared for by the next exercise.

The Sensible-Supersensible Plant

This exercise can be done with the concept 'plant', which leads us to the percept more concretely.
We presuppose experiences with plants and try to convert what they present to the senses into inner perception, in correspondence with the seed exercise in Rudolf Steiner's *How To Know Higher Worlds*.
We begin by looking at a seed and allowing a plant to grow out of it in our imagination. With the second step the starting point is a fully-grown plant in flower. We observe that closely too, and then imagine it wilting and passing over into the formation of new seeds or buds. Here too we should pay attention to the feelings arising from the inner activity, which are quite different from the ones we have when we allow a germinating seed to grow.

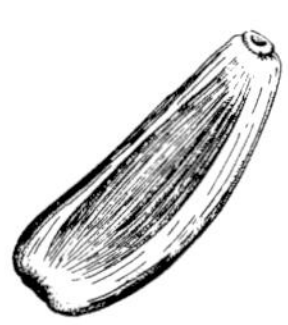

We place before us a small seed from a plant. Starting with this insignificant thing, the point will be to think the right thoughts intensively, and by means of these thoughts to develop certain feelings. First we must establish what we are really seeing with our eyes. We describe to ourselves the form, color and other properties of the seed. Then we ponder the thought: "This seed, if planted in the ground, will grow into a complex plant". We visualize the plant, we make it present to and in us. We build it up in imagination. Then we think: "What I now visualize in my imagination, forces of light and earth will later in reality draw forth from this small seed. But if this were an artificial seed, an artificial copy so perfect that my eyes could not distinguish it from a real seed, then no forces of earth and light would ever be able to draw forth such a plant from it". If we can clearly form this thought and bring it to life within us, then we will be able to form the next thought easily and with the right feeling: "Within the seed already lies concealed what – as the force of the whole plant – later grows out of it. The artificial copy of the seed has no such force. Yet, to my eyes, both seeds look the same. Therefore the real seed contains something invisible that is absent in the copy".
Thoughts and feelings should now focus on this invisible reality. We must imagine that this invisible force or reality will in the course of time change into the visible plant, whose color and form we will be able to see before us. We should hold the thought: "The invisible will become visible. If I were unable to think then what later becomes visible could not announce itself to me now".
. . . Meditative thoughts need to be experienced calmly and peacefully. No other thoughts should distract us . . . the thought, and the feeling united with it [should be allowed] to penetrate the soul. If this is done in the right way, then after a time – perhaps only after many unsuccessful attempts – we become conscious of a new force within us. This creates a new perception.

The seed seems to be enclosed in a small cloud of light. In a sensory-spiritual way, we sense it as a kind of a flame. At its center we experience a sensation similar to the impression made by the color purple, at its edges a sensation similar to the color blue.
What we could not see before now becomes apparent to us, created by the force of the thoughts and feelings that we have awakened within us. The plant – which is still physically invisible and will not become visible until later – is revealed to us in a spiritually visible manner.
Understandably, some people consider all this mere illusion. Many will say: "I do not want to have anything to do with such hallucinations and visions". Some will fall away and abandon the path. What it comes down to is this: fantasy and spiritual reality must not be confused in this complex matter of inner development. We must have the courage to push forward and not become faint-hearted and fearful. On the other hand, we must always cultivate a healthy sense for the distinction between truth and illusion. We should never lose conscious self control during the exercises. Our thinking must be as certain and reliable in carrying out the exercises as it is when we apply it to the things and processes of everyday life.
. . . It would be very bad indeed if we fell into daydreaming. Our reasoning must be perfectly lucid, even sober, at all times.

The extracts on pages 33, (right) 34, (left) and 36, (left) are from Rudolf Steiner, *Wie erlangt man Erkenntnisse der höheren Welten?* (GA 10), Dornach, 22nd edition 1975, p. 60; trans: *How To Know Higher Worlds* (Spring Valley: Anthroposophic Press, 1994) Ch. 2 , 'Controlling our Thoughts and Feelings' pp. 56–60

If I want to make an exact picture of a perceptible seed in the first step, I must enter a spatial relationship with it. I can think of the perceived object in outline. Of course, the *possibility* that a plant can grow out of it is not visible. It is spatially-temporally undetermined and it has no imaginable form. But it has a distinct spiritual 'direction' ('yarrow' for example) with which I orient myself on my path of allowing the plant to grow out of it in my mind. This acquires 'outlines' through temporary mental pictures of previously observed plants, *which should be remade* so that I can allow them to cohere into a particular, physically possible form.
To reach the germinating forces connected with a seed that could be encountered in space, it is important to experience the 'coloration' of this inner activity. Here Rudolf Steiner emphasizes that it is not a matter of hallucinations but of feelings that we can compare with the ones we have when we perceive the corresponding colors. They are clearly of a *soul nature*. I must inwardly take back again everything that I imagined in order to approach the inner perception, so that my gaze is directed at the mobile 'in between' and my experience can rest on it. In other words, what we are dealing with here is the plant's cosmic side.
Cosmic is used in the original Greek sense, i.e. it means everything we become aware of through perception of thought that is at the same time experience. Depending on our outlook, the stars, as usually conceived, are either objects or a special case in which the above can be experienced.

A further exercise, connected to the seed meditation, is the following. We place before us a mature plant. First, we immerse ourselves in the thought: "A time will come when this plant will wither and decay. Everything I see now will then no longer exist. But the plant will have produced seeds, and these will become new plants. Thus once again I become aware that something I cannot see lies hidden in what I can see". We saturate ourselves with the thought: "The plant form with all its colors will soon no longer be there. But the knowledge that the plant produces seeds teaches me that it will not disappear into nothingness. I cannot see what preserves the plant from disappearance anymore than I could see the future plant in the seed. Therefore it follows that there is something in the plant, too, that I cannot see with my eyes. But if I let this thought live within me, and the appropriate feeling unites with it, then after a time new force will grow in my soul and become a new perception". A kind of spiritual flame form will then grow out of the plant. Of course, this flame will be correspondingly larger than the one described in the case of the seed. It will be felt as green-blue at its center and as yellow-red at its periphery.

Again, it must be strongly emphasized that we do not *see what are here called 'colors' in the same way that we see colors with our physical eyes. Rather, through spiritual perception we experience something similar to the impression made by physical colors. To perceive 'blue' spiritually is to feel or sense an impression similar to the one we feel when our physical eyes dwell on the color blue. We must remember this if we wish to ascend gradually to true spiritual perception. If we do not, we will expect the spiritual world to be a replica of the physical world, which would confuse us in the worst way.*

Ibid pp. 59–60

In the transition from the fully-grown plant *observed with the senses* in the second step to seeds and buds, we now have to do the reverse, i.e. the mind must let go of the differentiated, visible phenomenal picture. Just as the forms of summer foliage retained for a while in the autumn gradually disappear and give rise to humus, so a separation from the completed form is carried out inwardly. Imagined pictures as such cannot be undone. But a capacity arose through the activity of allowing the mental image to arise in connection with the percept, which now helps us to make a conscious transition to an inner perception and experience of the plant's formative possibilities. In this way we have an experience that corresponds to the general fertility of the humus that arises from leaves.

The process that we experience in the physical formation of a seed does not, as in the case of withering leaves, fall out of the process of formation of the individual plant – it continues in it. A form (fruit) that is complementary to the fully-grown plant arises from what has not yet taken on a form in the plant. This is complementary to the extent that it is oriented inward towards the formative capacity, and not outward, towards differentiation and dying – as we see most clearly in the corolla. As with going through infinity in the projective geometry exercise, here a new, imaginable thing arises with the seed, which however seems to be spiritually inverted in the opposite direction. Instead of 'fertile humus' a seed arises as a separate thing with new formative possibilities, but set in the 'spiritual direction' of the plant species.

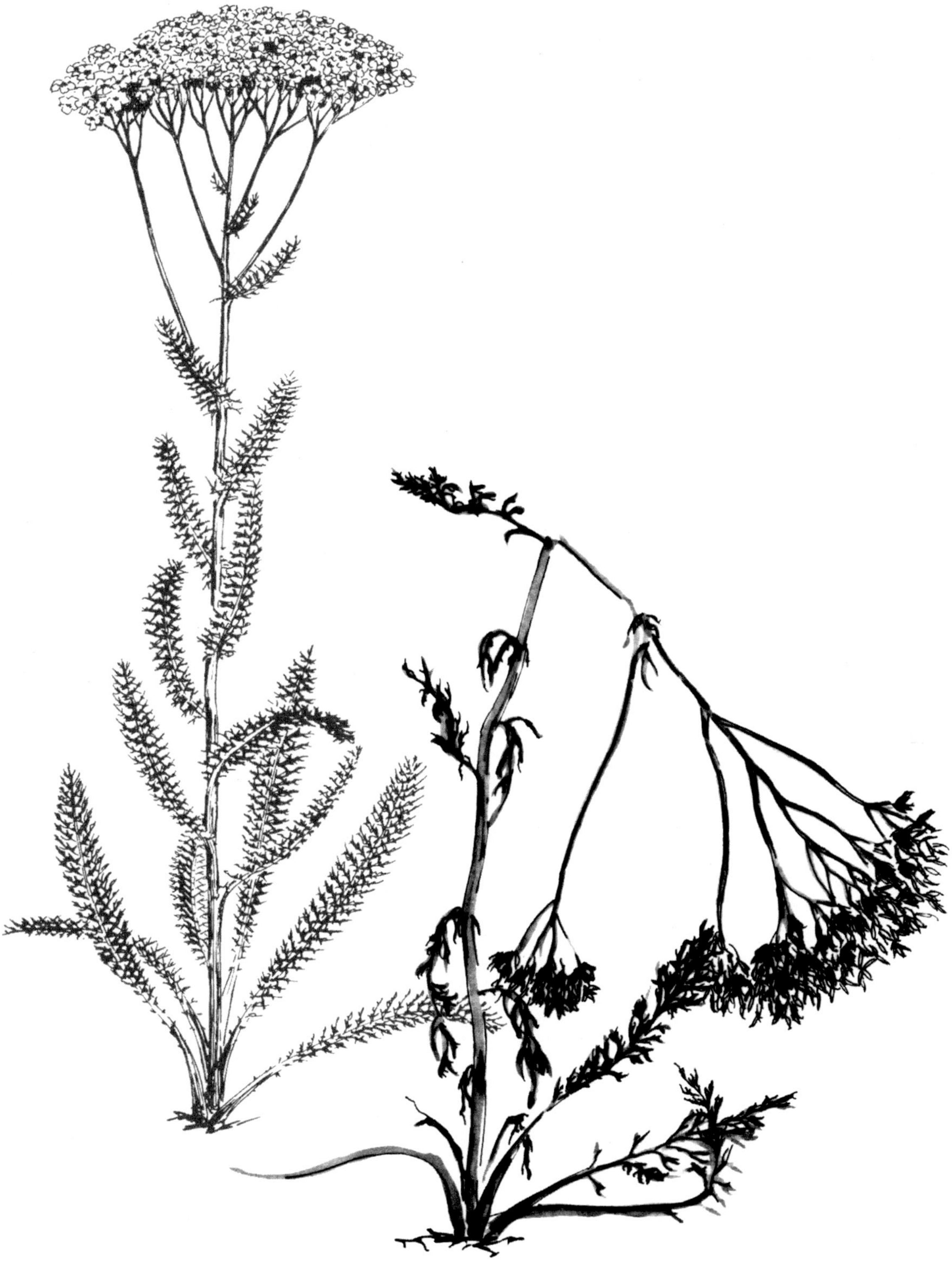

The preceding exercises help us to perceive things with our thinking. This is the stage of imaginative cognition – a first step towards overcoming the subject-object boundary that only exists in objective consciousness. Goethe's work with the metamorphosis of plants can be looked upon as a first attempt to actively grasp the connection between the way plants form and the 'type' idea. People often see only the outer side of this, i.e. efforts to arrive at a factual, morphological classification of plants. But they are less aware of Goethe's special *way of experiencing* and they do not realize that it enables inner access to the nature of living things. A general look at a plant's growth and decay, as indicated in the previous chapter, has already shown us that we can only approach the reality of life if we make a conscious transition from sense perception to formative capacity and an inversion into the new phenomenal form. With his views on metamorphosis Goethe was the first to try this. Rudolf Steiner called this undertaking 'grasping the spiritual with the soul'. If we follow Goethe's example of the exact observation of a plant along the lines of the previous exercise, at first we have in mind an indistinct perception of the plant and we use this perception in formative, mobile, imaginative thinking to orientate ourselves.

Rudolf Steiner, 'Goethes Metamorphosegedanke' in Die Grundimpulse des weltgeschichtlichen Werdens (GA 216), Dornach, 3rd edition 1988; trans: 'Goethe's Metamorphosis Idea' in Supersensible Influences (London: Rudolf Steiner Publishing, 1956) Lecture 5, 30th Sept. 1922

Five developmental stages of a nipplewort, *Lapsana communis* L. Lots of turning points occur in a plant's developmental process which we experience when we run through it afterwards. In the example shown here: 1–3 spreading of the rosette. The plant is mainly oriented towards root formation in the earth.

4. The shoot stretches upwards. A contraction becomes visible in its sequence of leaves.
5. Now the side-shoots have extended from the leaf axils. Visible growth is completed. The yellow flower stars radiate into the surroundings.

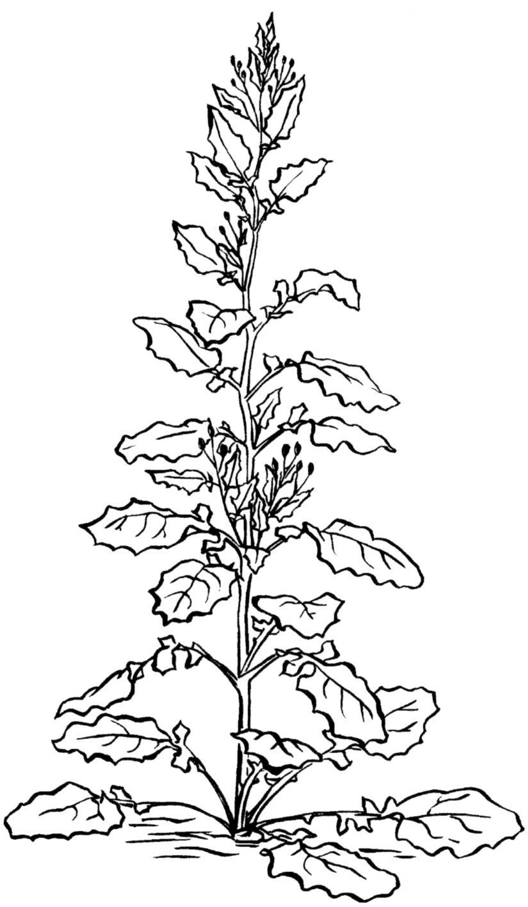

Seven stages in a growing nipplewort leaf. The line drawings below indicate the correct relative sizes.

Seven leaves of a nipplewort. Below the complete leaf sequence in correct size relations.

One only becomes familiar with this perception by working with it. Through ever-new percepts it becomes separated from one-sided connections, expands and increasingly matches reality. In this way the initial provisional perception serves as a tool for taking hold of imaginative cognition. Through our special contemplation we become aware of what we do continuously in everyday life, namely, convert sense experience into thinking experience and this in turn we change into something that is accessible to outer experience in words, pictures and deeds.
Connecting with the life of a particular plant like this means starting to build a bridge from a knowledge of material plant-formation to an inner experience of our own organization.

Metamorphosis Experiences with Growing, Greening Plants

Every step on the path described here opens up new experiences. First we grasp the reality of the plant we are faced with. We become creatively active inwardly when we begin to follow a growth process or study a leaf series. A percept arises through our own activity. Here our alert 'objective consciousness' becomes an inner observer, a controller who tells us each time what relationship our inner activity has to the outer world. For instance, if we study a *growth process* throughout its stages we connect percepts that we made in succession. But in principle we could look at it repeatedly and in doing so come across intermediate stages. Our creative activity grasps something that takes place in the sense world. This is not the case when we look at a *leaf series*. Here we are also creatively active inwardly. Admittedly the stem gives us a spatial, temporal relationship between the leaves. But the *transformation* that we see in the succession of forms does not correspond to any outer situation as it does in the growth of a single leaf. For the successive leaves that arise from the meristem are always new. We only see the development of a plant through the interaction of the two formative movements.

◁
We also discover turning points in the course of the development of a single leaf. First of all a narrow tip arises out of the meristem in a bud. Next it begins to divide: a blade and stem are added. This is all still hidden in the bud. It is only in the spreading of the blade and the extension of the stem that the leaf begins to relate to its surroundings.

◁
The change in form in the series of completed leaves goes in the opposite direction. It begins with a stemmed leaf. Increasingly distinct indentation follows. The stem and blade gradually fuse during the contraction phase and form a common tip.

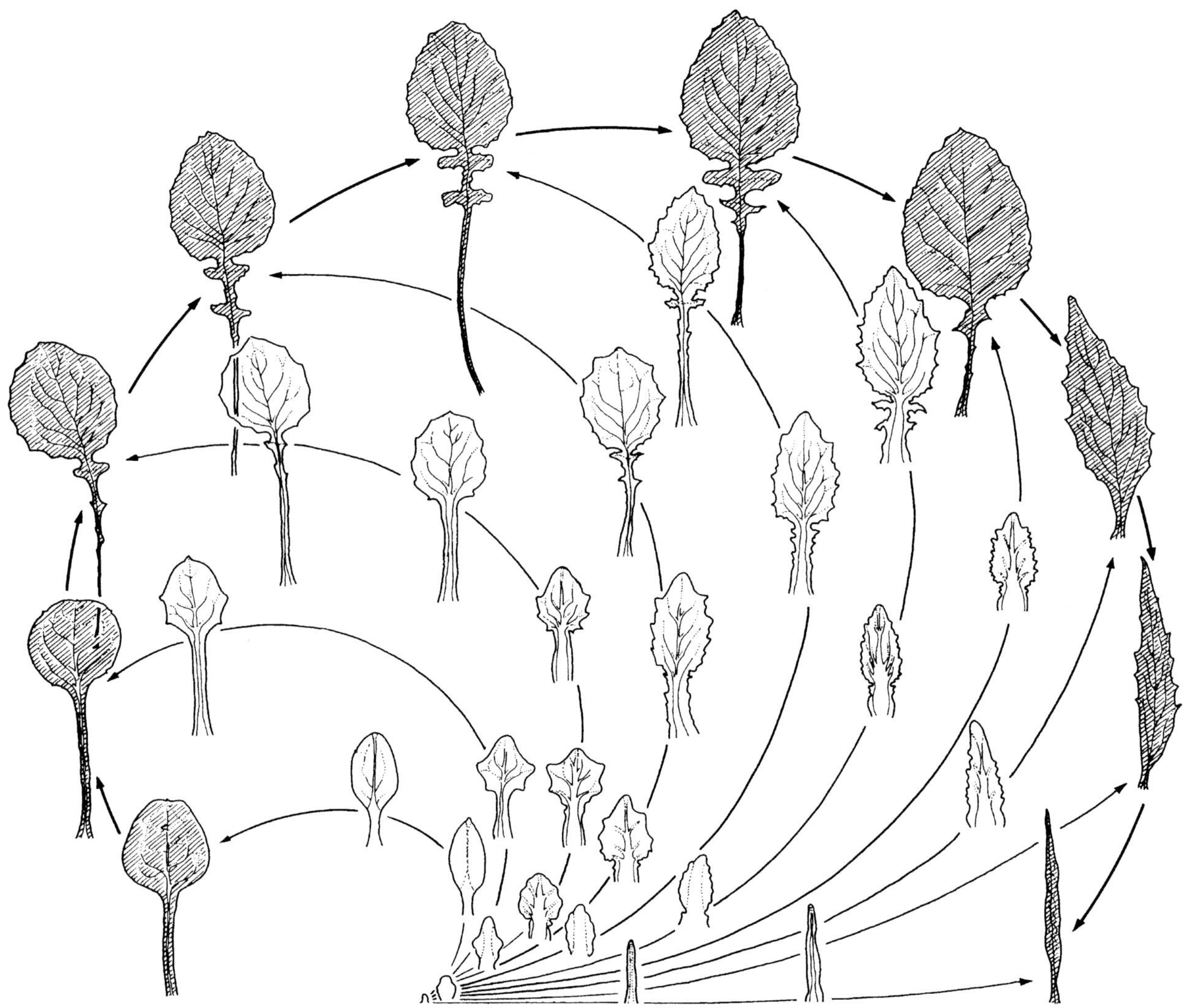

Overview of the development of single leaves in a series of nipplewort leaves. The fully-grown leaves are shaded. The way individual leaves are formed (anti-clockwise curves) matches the changes in the end forms (clockwise outer curve) in a way that shows that a time organism changes continuously and yet can remain a whole.

If we open our eyes to the process we then grasp the plant's formative principle from within and participate in its life. We experience in increasing detail how the plant continuously passes over from an invisible form to a visible one, not just on the way from seed to fully grown plant and back to seeds, but in the formation of every single organ. It reaches a boundary and at the next leaf the formative process begins again in a slightly changed form. But this also means that a change takes place on the invisible side of the shoot ('inside' it).

Thus something happens at the formation of the first leaf. Something immanent within the meristem takes on outlines as it grows. In this way inner conditions for the next leaf arise. Imprinted on the organism is something like a capacity that only expresses itself in the next step in more differentiated forms. Even in ferns there is a kind of metamorphosis. It comes to expression in a progressive differentiation of forms. We get the impression that this process could continue endlessly.

But there is less differentiation in the formation of further leaves. If the plant does not flower there seem to be only repetitions which change in accordance with environmental conditions and do not show any progress (the forms could also become simpler if the living conditions become more favorable).

South American fern
Pteris sp.

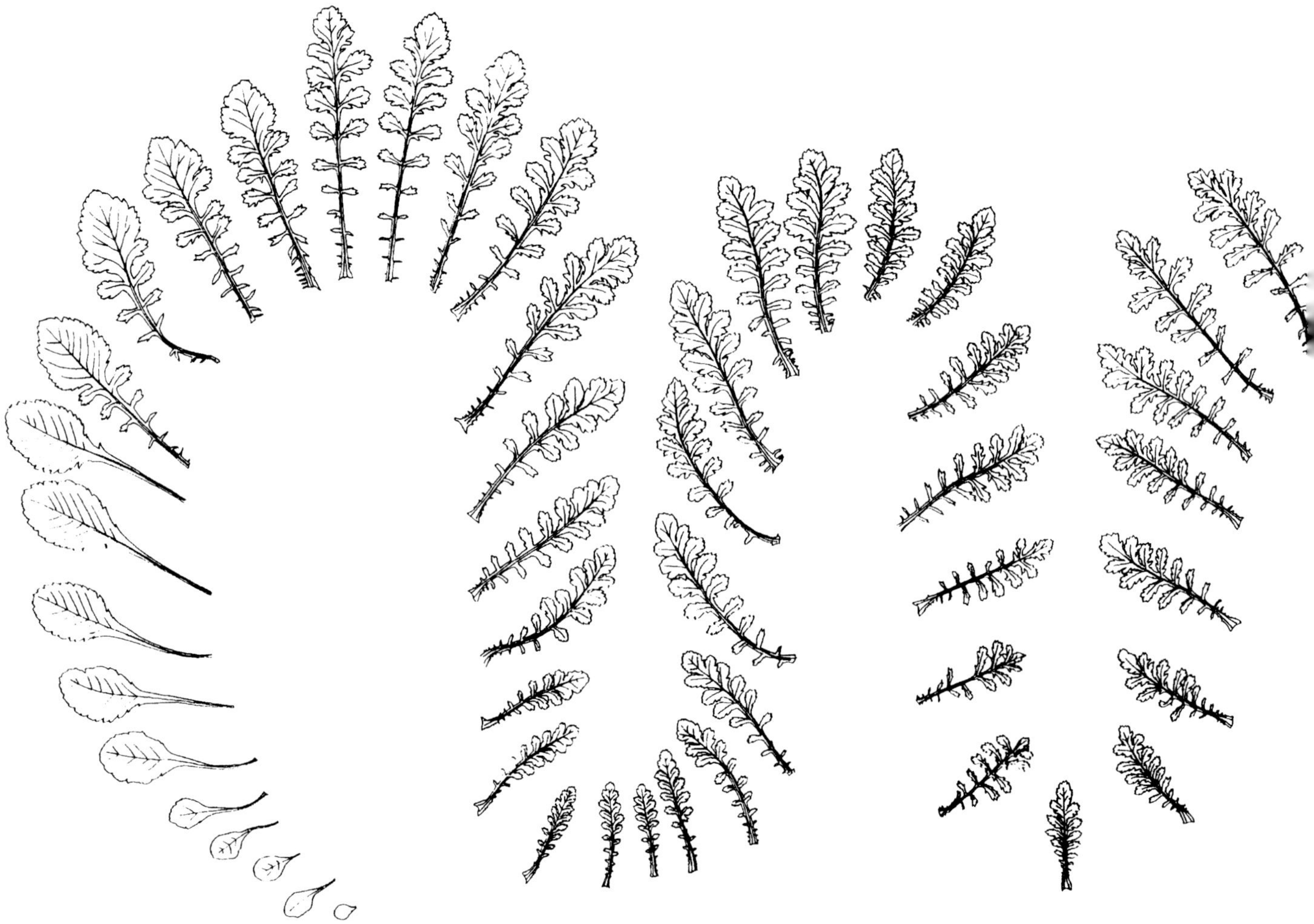

1st year *2nd year*

This series of scabious leaves shows that more or less the same thing is repeated over several years: expansion of rosette leaves towards summer and a contraction towards winter. But when the stem extends and development proceeds to flowering the contraction takes on a different character: we notice an increasing tendency towards differentiation, independence in the pinnation and at the same time, a simplification.

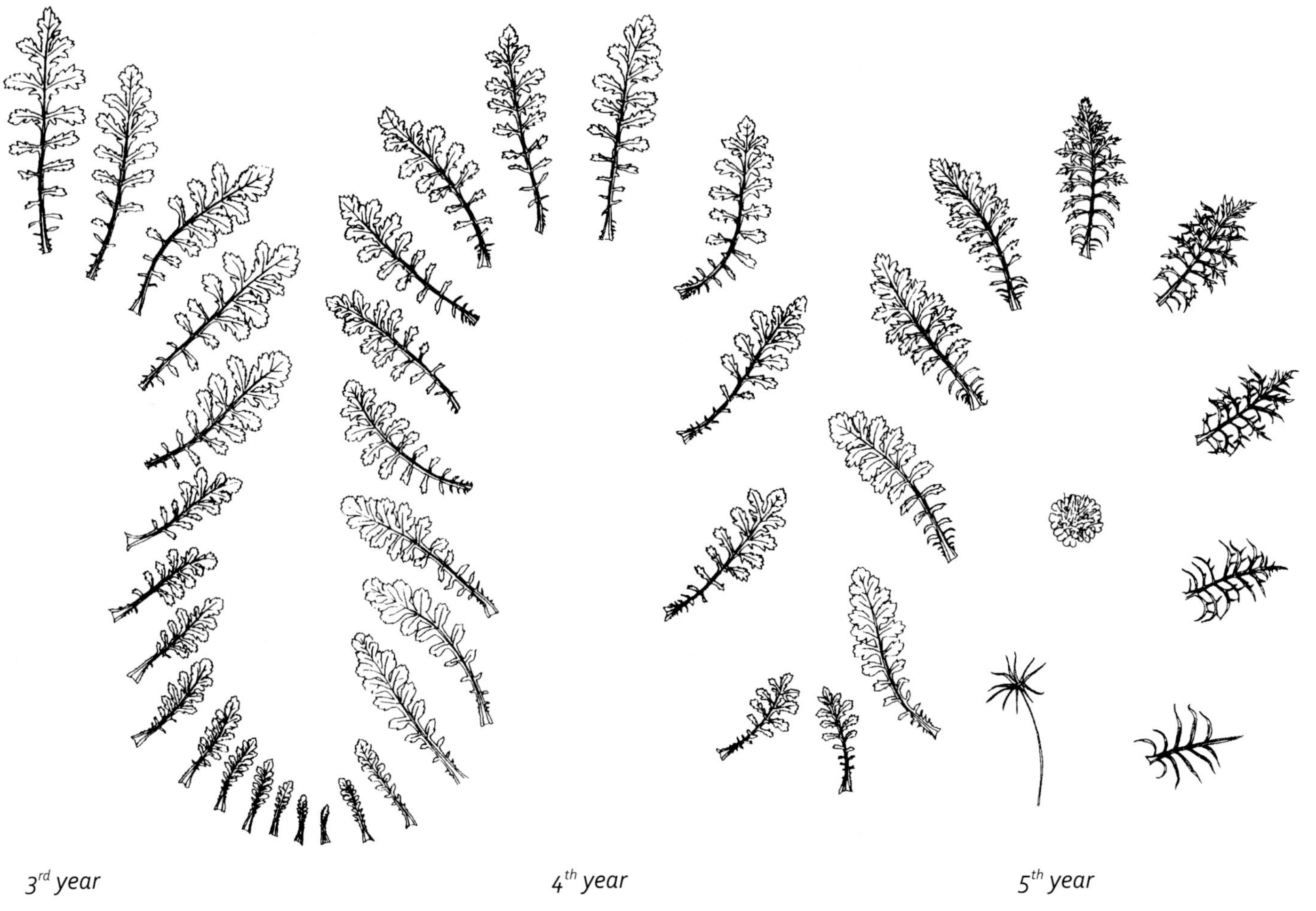

Leaves of a scabious, *Scabiosa lucida Vill.*, from a seedling to the flower five years later.

But even in ferns something often changes when *spore-bearing leaves* are formed. These appear to be more contracted than the previous ones. They are often slightly rolled at the edges and sometimes are even like an independent flower organ.

The formative possibilities of the whole plant decrease from node to node towards the flower. The closer a side-shoot – which springs from the eye (bud) in a leaf axil – is to a flower, the more suddenly it passes over into flower formation and the less directly it can develop a new plant that grows out to the side.

Moonwort *Botrychium lunaria* L. Leaf and sporophyll

▷

Nipplewort: main shoot and six primary side-shoots from below upward, with the corresponding series of leaves.

Sets of five leaves of the leaf sequences of nipplewort plants from various localities. Top: poorly developed plant on poor soil. Bottom: luxuriant plant on good soil. Transitions in between.
Very different plants arose, but even the highly simplified stunted form became a flowering and fruiting plant.

In doing this we have approached a plant's mode of formation only in a one-sided way. We connect what we see with what we do not see, but which we look for in inner perception as a complement that is demanded by the object under observation. While it is true that the complement we have been talking about, which is experienced inwardly through changes in forms, is a kind of 'spiritual environment' for sensory things, we relate this environment, in a way that accords with the facts, to what from the past is connected with the substances of a plant (heredity). Indeed, the plant could not really develop without an environment that is active in the actual sense world.
Therefore it is necessary to look at another aspect that plant formation points to, namely light, warmth, water, soil conditions, the overall atmosphere of the place where the plant is growing, seasonal changes, and so on. We can see their actual connection with the plant only as an 'effect' in the picture of the plant and not directly.
Here, too, it is a matter of letting the appropriate surroundings arise in inner perception. This is another kind of inner complementation of the way forms metamorphose.
The views about formative receptivity and formative activity correspond to the *moon* and *sun* in the cosmos: the moon does not itself produce light. We observe that it is a 'material object' that has a spatial, rhythmically moving relationship to the stars. We experience daily how the *sun* activates all terrestrial life with light, rhythm and warmth. An individualized reflection of sunlight arises through the changing receptivity of the moon, which works mainly on the night aspect of existence.
The metamorphoses of forms are accompanied by changes that we notice through color, smell, taste and touch. We need only consider the freshness or witheredness of plants outdoors or indoors; how pictures and connected experiences arise in our consciousness when we *smell* things and how *tasting* brings us close to effects in our body.

This brings us to a distinction: although the vivid contemplation of plant development that was gained on the path first described is indeed connected with substances, it does not include the forces that actually produce things, although it does let a perception of the receptivity for these arise. On the other hand, if we seek the mode of formation as something resulting from the effects of the sensorially experienceable environment, we become familiar with the productive forces for which the plant's substance is receptive. This receptivity is connected with the physical germ and is subject to heredity. The holistic (etheric) forces of the surroundings (such as warmth, light, rhythm, conditions in the overall situation) can become formatively active so long as the plant, stimulated by the elements (warmth, air, moisture, earth), is placed in the condition that is appropriate for it at the time, for example, the swelling of the seed. We see that this activity generally comes from the sun. We are dealing with the etheric, with the growth-promoting aspect of the sun's activity. But the effect of the sun varies in a specific way according to its position, i.e. its position in relation to the earth and the zodiac. It mediates a special 'constellation' of the cosmos. This is the astral aspect of the sun's effect that determines the actual mode of formation, limits growth, relates to the special character of a plant species, and also makes it manifest – compare shade loving plants and those preferring direct sunlight.

Three stages in the development of columbine (*Aquilegia vulgaris* L.):
1 before flowering;
2 flowering: the bluish-violet catches the light; in the stand they can now be seen everywhere;
3 after flowering: the green returns during fruit formation; the columbines disappear again into the general greenness of the stand.

Contemplating a Flower as a Condensed Experience of Metamorphosis

In such a sequence of exercises, what arises in consciousness is something like a condensation. It forms a spiritual space. In it all of the plant's formative possibilities are present simultaneously, 'intermingled' as an experience of the whole, as a thinking perception that is permeated in varying degrees with experience that had previously been obtained in sequences of events. I do not have to go anywhere for this, for it is as if I were inside the plant. I am contained in this view and I face it inwardly (inspirationally) as a part of my being whereas I face the physical world outwardly. The reality of a living plant includes both of these.
We cannot imagine this overall perception. The important thing is the force from which the mental images that induce us to say 'yarrow' are created. Present in this force is what becomes ever richer inwardly with each encounter with a sensory percept. On the way from seed to fully developed plant and back to seed we discover in our imaginative reconstruction that the plant is a being in the sense world that exemplifies what we learned to handle through the preceding exercises through our own connections with the world.
In growth: diving down into the sense world; becoming a *one-sided* phenomenon by comparison with an (etheric) 'universality' in space that is active in the sense world but that can only be grasped in thinking experience.
In withering: withdrawal from the sense world; becoming a memory that can reverberate in our consciousness. The germs of a new openness for sensory phenomena form at the same time.
One can follow the metamorphosis of leaves right into the flowers and consolidate it into a complete view of the course of a plant's life. Indeed, the upper leaves usually remain green, i.e. alive, up to the end of fruit and seed formation. A plant constitutes a whole through the fact that earlier conditions are

Flowers and leaves of the columbine, *Aquilegia vulgaris* L. Below: Selected leaves from the main shoot (see p. 54 for complete series)

The greening process can be looked upon as a whole. It spans the whole developmental process from germinating seeds to fruit formation. This is already indicated in the expansion and contraction in the leaf series. The flower seems like a brief recapitulation at a higher level and a gesture appears.

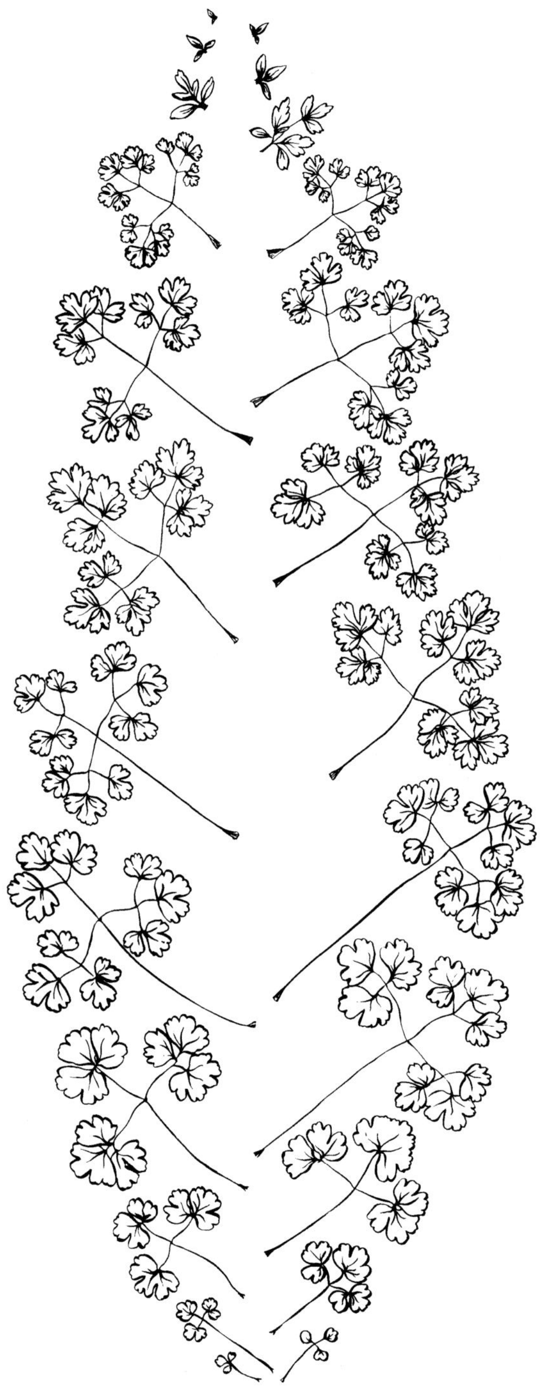

Complete leaf series of columbine

carried over into later ones. In this way we see a gesture and can get a feeling for the course of the interaction between the plant and its environment, and in it, a sense of the character of the leaf formation and the integration of the plant into the course of the year.

In contrast to this, a flower is a completely new phenomenon. It is there all at once. Let us try to compare what we experience in a flower with what we feel from the leaf process in the way outlined. We could describe the difference as follows. It is true that a flower presupposes growth processes, leaf development, and all vegetative processes. But these are withdrawn when the flower opens. Something entirely new appears – an overall shape that no longer has any outer similarity with leaf shapes, an overall picture whose colors (usually not green) are fully integrated with the shape and around which spreads a more or less distinct, specific smell.

Now the important thing is to compare the gesture experienced in the overall form of the green plant with the flower and to let the two different kinds of expression illumine each other.

The outer connection is given through the fact that leaf and flower grow on the same stem. The inner one is given since we are dealing with a yarrow or a columbine, and so on.

We can no longer understand the flower's shape in connection with receptivity and the actual effect of the elemental environment, i.e. viewed cosmically from the interaction of sun and moon. Flower shape in any one plant species is always the same and seems to be independent of these effects. We can experience flowering and the corresponding form motifs of the whole plant as gestures. Like concepts, gestures are only comprehensible through their context. Here the context is the manifold shapes of the entire plant kingdom. Similarly, we do not recognize Sirius in the night sky if we just look at the single point of light, but only after we have related it to Orion and the starry cosmos.

Fruit and Seed

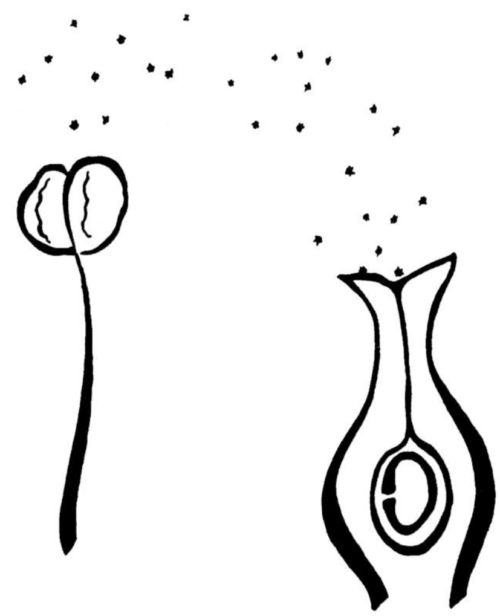

Only a combination of both the inner experience that is suggested by the leaf region and the picture of the flower that is turned outward give a view of what ripens in fruit and seed. What arises in our consciousness becomes seminal and contains a tendency that is inherent in the substance of seeds and out of which new things can arise – or nourishment that, as transformable substances, represents the basis for new forms.
The two poles (ovule and pollen) that separated during flower formation are brought together substantially in the transition from flower to seed – the ovule as an *'internalized' vegetative capacity*, i.e. as a bearer of receptivity, and the pollen as a *'contoured' vegetative capacity* that gives an impulse for a new formation determined by cosmos and earth.

Root Formation

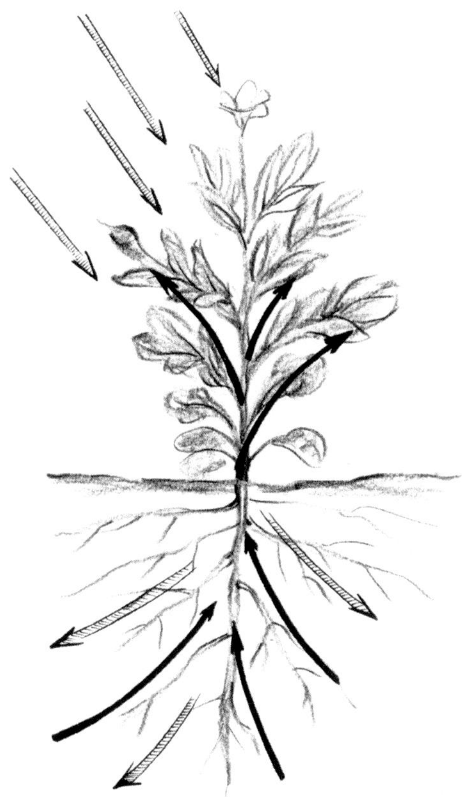

The roots are often not included in such an exercise. They are usually not observable anyway. But in medicine they play a very important role. For this reason it is especially important to include them in this guide. When we study root development we have quite different experiences from those we have when watching the formation of organs above ground, for roots do not put out leaves, flowers and fruit. On the contrary, a root inserts itself into the dark, firm earth environment as it goes in various directions – either downwards following gravity, or in several directions towards aerated water. So what we notice above all in a root is its activity of growing into its environment – pushing its quick-rotting cap in front of it as well as secreting juices and absorbing dissolved substances (mainly minerals). What the plant, stimulated by warmth and light that works via the leaves, initially 'experiences' in the darkness of the earth then takes form from below upward in its organs above ground. Through this process it acquires the ability to form

▷

Rhizome and root of angelica, *Angelica archangelica* L. (left, cut open). The rhizome retains a shoot character. It has no shoots that originate from the roots.

▷

Rhizome and roots of herb bennet, *Geum urbanum* L. (left: freshly cut open, dark red inside). The shoot that continues to grow is repeatedly pulled into the earth as a rhizome by roots originating from the shoot, whereas the main root and lower parts of the rhizome die away.

and solidify itself. Thus a complete view of a plant includes the inversion in the breathing interaction between processes above and below ground.

General Principle of the Main Axis

Anyone who has worked more closely with the preceding plant exercises may have noticed that an invisible connection becomes perceptible throughout a plant's developmental steps. We discover this if we selflessly perform the transformations in our own minds to the extent that we identify ourselves with them. The main axis from root to shoot and into the new seeds is the visible trail of this experience. All the processes of formation and transformation of plant substances ultimately proceed from this axis or come together in it. From a cosmic viewpoint this involves the most comprehensive part of plant formation that can be grasped in our own self, namely a specific relation of the earth to the sun at any particular time, which in essence points to the original unity of the cosmos.

Against this background the visible plant becomes a picture for us that proceeds from the activity of the cosmos. This picture can make us aware in various specific ways of how we develop activities at different levels in our body. In the metamorphosis of our own will, from life in the limbs via life in the rhythmic organs to life in the senses and thinking, we become aware, through our identification with the corresponding gestures, of the reality of the will that is active in the world as substance.

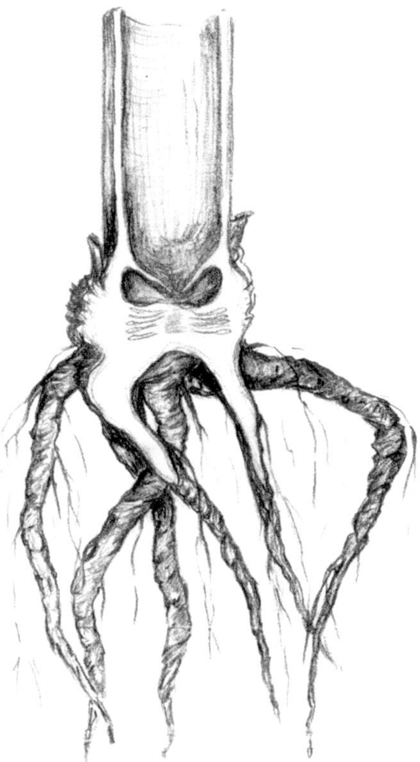

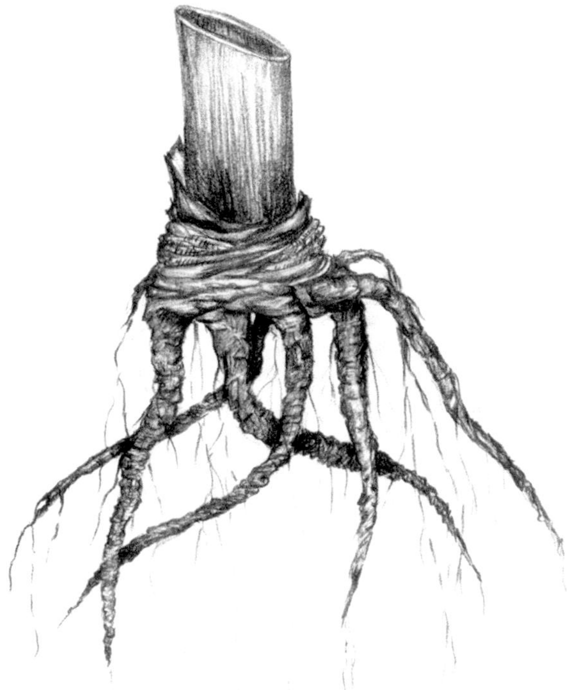

Rudolf Steiner, *Das Verhältnis der verschiedenen naturwissenschaftlichen Gebiete zur Astronomie* (GA 323), Dornach, 2nd edition 1983. (Translated by George Adams, typescript available only)

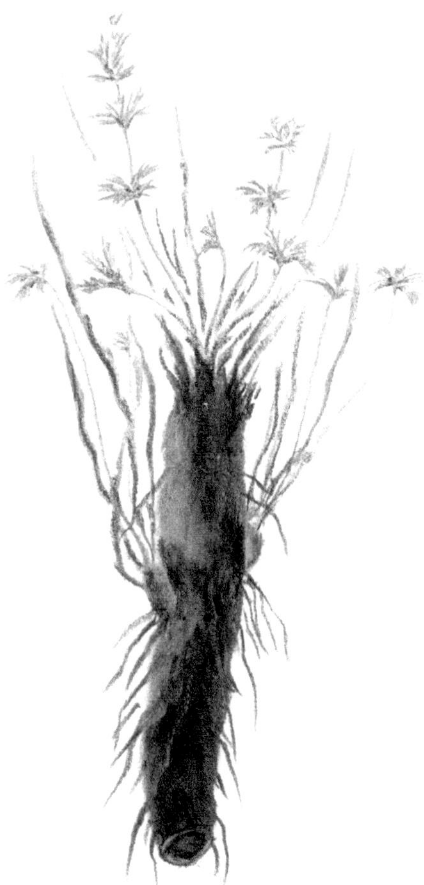

Rhizome of tormentil, *Potentilla erecta* L. Freshly cut places are brownish-red. The rhizome becomes more like a turnip.

In his third scientific course, Rudolf Steiner set scientists and mathematicians the task of transforming human limb bones into skull bones with the aid of projective geometry. This task takes on a greater significance against the background of what was just said, if the orientation for the corresponding inversion is taken from self-experience, i.e. from the observation of the way we experience that we are active in our organs. Only then do the solid forms of our body become intuitively graspable pictures for the activity of the I.

Hence thinking about the plant's main axis, its way of transforming, and the way that it follows directions in space in connection with processes in its various organ regions is particularly important for an understanding of healing substances.

From this viewpoint we can properly understand in detail what the 'displacements' we discover in plants means for the nature of substances. What seems to be the same substance can actually be something different, depending on whether something 'flowery' appears in a flower or in another organ, for example, the root.

We should pay particular attention to transition zones, especially between root and shoot, and their manifold interpenetrations, as in rhizome formation. It is indicative that many healing plants are called 'wort' or root, such as snakeroot, St. John's wort, woundwort and so on.

Imaginative Contemplation of a Plant

Through the capacity developed by the exercises described we begin to link inner to outer in steps (perceiving, feeling, forming and transforming mental images and observing this activity); we start creating context, i.e. getting to know how to deal with etheric forces in our thinking and thus perceiving our own etheric or life body. We learn to experience how our etheric body is active in particular situations through handling these processes when observing our thinking. This activity is conceptually shadowy at first but it can gradually be experienced as something that is 'full of content and concretely pictorial'. So we follow the life of a plant imaginatively and gain a perception of its etheric body with the aid of the consciousness in our own etheric body. We are not postulating a life force but instead we are experiencing it as something real that is perceived in vivid thought experience. This imaginative experience of a plant makes us aware of the plant's receptivity for particular environmental conditions (moon) and the peripheral forces (sun) that work inwards and create actual connections. So we do not look upon plant form as something that is formed causally by physical, material principles, but as the result (effect) of these two cosmic sides of the etheric body and earthly substances. We encounter the substances in heaviness and in touching (also in that we 'touch' things with other senses) and the etherically active forces through what is of a pictorial nature in form, seeing, smelling and tasting.

Flowers and the whole form speak to our soul. This touches our astral body and it is the pictorial beginning of inspirational cognition.

What proceeds from the main axis in the course of reconstructing it inwardly can eventually lead us to a *threefold experience of the I* which then appears in the seed as a fourth experience that encompasses everything. We will go into this further in the next chapter in connection with higher cognitional capacities.

Perceptions gained through enhanced powers of thinking are far from pale and shadowy; they are full of content, utterly real images; their reality is much more intense than is found in the content of our sensory impressions. A new world opens up for human beings when they have extended their powers of perception in the indicated way.

Rudolf Steiner and Ita Wegman, *Grundlegendes für eine Erweiterung der Heilkunst nach geisteswissenschaftlichen Erkenntnissen* (GA 27), Dornach, 5th edition 1977, Ch. 1, p. 7; trans: *Fundamentals of Therapy* (London: Rudolf Steiner Press, 1996) Ch. 1, p. 4

Physical Matter, Mineral Substances, Chemical Properties

If we work with live plants in the way indicated in the previous chapter they become soul content and we gain a much more comprehensive, mobile concept and a grasp through the soul of the etheric-astral formative forces that create pictures in what is sensorially given. Thus the special directions of influence of formative, motive forces in matter become experienceable.

At this stage we find clear distinctions between matter and substances as well as between physical mineral substances and live ones.

Any mineral or plant that we respond to sensorially and about which we are disposed to reach judgments is physical. So are substances such as starch, salt, and air. Their nature is described by form, color and composition. Their *chemical properties* only become perceptible when the substances are dissolved in water, or they lose their previous nature in some other way, and perhaps reappear in the form of new phenomena.

The sequences of phenomena arise in the interrelationships of substances, for which we create the conditions in chemical experimentation. Thus we establish their potential for action. Looking at the transitions from one phenomenon that is given under certain conditions, to another with changed conditions, already requires an imaginative way of looking at things. We do not ordinarily make this very clear to ourselves. When doing chemistry we do not encounter the reactant as matter, i.e. as 'objective', but we experience the substance in the way it changes under the given conditions or, in other words, in the process and not in fixed atomic models with which we later connect our experiences abstractly in order to grasp them conceptually.

The various phenomenal forms of substances receive a meaningful context through the way they occur in nature. If we want them to arise by producing the physical conditions for them ourselves, we

Salt cube

go to a suitable place and either copy nature or, to the extent that we are dealing with phenomenal forms that had never been perceived before, create new nature.

Matter becomes *substance* for us the more we have in mind the present, sensorial, phenomenal form of the material and its ability to produce effects in a natural context. We know that silica, calcium carbonate and sodium chloride are *minerals*, for we observe their modes of occurrence such as rock, mountain and ocean formation in the context of the whole earth. We can look at carbon, hydrogen, nitrogen and oxygen in a similar way.

We see the physical substances of plants such as starch and essential oils arising from plant formation. They can become similar to mineral substances and, as with crystalline sugar, take on purely mineral properties. These 'minerals' are distinguished from others through their origin and the particular way they are produced which remains more or less attached to them (cane sugar or beet sugar for example) or through what we can or want to do with them. To isolate something chemically means to take away the context of the particular way in which it came into being.

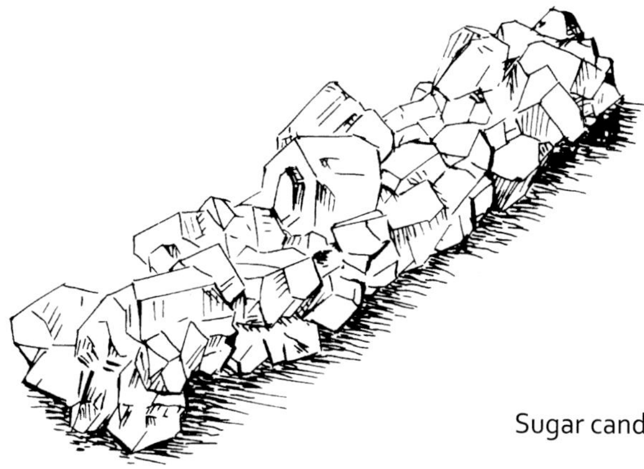

Sugar candy

Absorption of Mineral Substances by Living Plants and Creation of Plant Substances. Ascending and Descending Streams of Substance Formation

From a *physical* viewpoint, a plant lives by taking in mineral substances and in doing so grows and changes progressively. The physical substances that arise in this way are an expression of the plant and, only indirectly, of the absorbed minerals. The latter substances can be recovered through chemical conversions. *Living substance* lies in the transition from one state to the other. It can only be grasped imaginatively through something like the exercises described in Chapter III.

From this viewpoint, what we read from the way the plant manifests conveys to us its flowing, varying

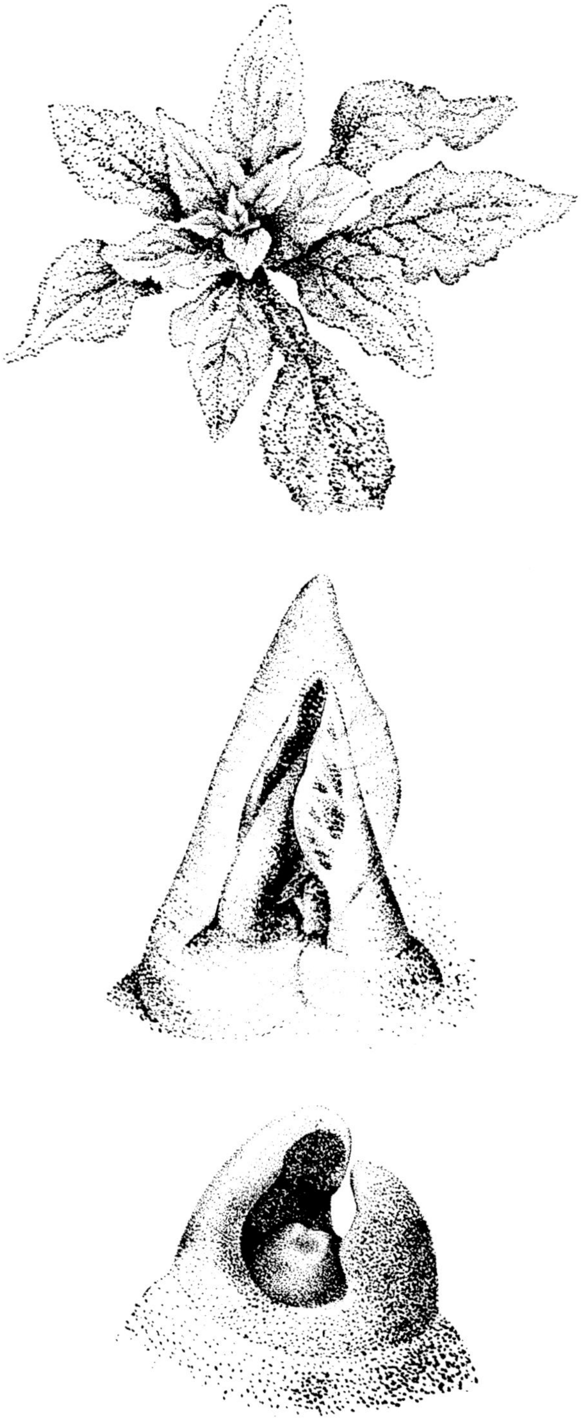

Rosette, youngest leaves in the heart of the rosette and growing point of foxglove *Digitalis purpurea* L.

connection with, on the one hand, etheric effects in the surroundings (light, seasonal changes, cosmic-terrestrial context of life) and, on the other hand, the elements (gravity and the consolidating, delimiting tendency of the earth; moist, fluid, changeable water; transparent, mediating air and warmth that brings forth everything).

In the activities of living substance, that we experience imaginatively through sensory phenomena, interact changing influences from the environment and changing receptivity for environmental effects and they form a continually varying whole. Thus changes in plant form proceed from germination to the formation of new seeds through what is earthly-physical interacting with what is cosmic-etheric.

Substances that are initially mineral also go through these transformations in the plant. We can reconstruct the route they take in a living plant and thence in physical plant substances again through imaginative perception by asking ourselves, "What capacity does a plant acquire through taking part in what in the broader context of nature we know as a particular substance, say nitrogen or carbon?" For instance, we know that there is a connection between nitrogen and a plant swelling and becoming bluish-green, or between carbon and the ability to take on forms. These materials in a plant gradually relinquish their mineral nature, i.e. their direct connection with a natural context (their 'primary effect'). They go into plant formation and thus now have an indirect connection with the sense world in the qualities of living substance and with *aspects* of a live plant. Thus we can speak of a 'carbon aspect', 'nitrogen aspect', and so on.

In this way a more general understanding of the processes of appearance and disappearance of plant form leads to a picture of ascending and descending streams in whose interaction a plant lives.

Ascending stream: in plant formation mineral substances are taken up by the roots (salts and water) and leaves (carbonic acid) into a realm where, under the influence of light and warmth, they completely

relinquish their mineral nature. The *plant* shapes itself out of them. It has the greatest *potential to alter its form* at its delicate meristem, which is constantly renewing itself, but is the least formed outwardly. Like a dewdrop as transparent as glass, the protein there, the physical aspect of plant substance, achieves its highest stage of vitality relatively far removed from earth conditions. This means that a plant is most mobile at its meristem. This point is the most sensitive and receptive one for the environmental effects that are mediated by the roots, leaves, and so on. On the other hand, to the observer, the plant's connection with etheric and astral forces results not from their perception of the context of appearance of its forms or other sensory qualities or material properties but, as in the case of seeds, from a living inner perception of the special nature of the plant. We become aware of the properties of plant living substance, from which its physical substances arise, as a boundary experience between encounters that are either pure perception or pure consciousness. The protein is obviously made in such a way that ever-new forms can arise from it in accordance with the characteristics of the species. Although it is physically present as a substance it can 'leave room' for the creation of new forms in a species-specific manner. Thus, standing opposite the plasticity of outer things, is a clear orientation towards the spiritual aspect of the species.
These formative possibilities become increasingly limited in the course of the plant's development, for example, when only flowers can arise from the meristems in the axils of upper leaves and not whole plants.

Foxglove in flower *Digitalis purpurea* L.

Developmental stages of a *Vitis vinifera* L. leaf

Descending stream: when a plant shapes itself spatially from the meristem and becomes differentiated as substance, its appearance is the result of etheric activities and it increasingly loses its mobility in this process. At the same time, the particular features of the species become increasingly visible to the senses. Differentiated earthly matter arises from which proceed new primary effects (in the sense of 'active principles'), but now with a particular connection with processes previously undergone.

One can distinguish stages of development in the formation of a leaf:

- Something new burgeons forth from what is outwardly undetermined, unfurls out of the meristem (impulse).
- A new form is begun, usually still in the bud and independent of spatial, environmental effects. It follows on from what precedes it, but its further development is still open (outline).
- The leaf emerges from the bud, spreads out, grows and adapts itself to environmental effects, i.e. orients itself in a specific way in light and gravity (spreading).
- A final form, which arises from the interaction of the outline with environmental influences, is completed. Substance synthesis (assimilation) continues (form completion).
- The whole process is the visible side of substance formation, synthesis and shaping. Whereas parts of the plant begin to decompose, substances develop with which a new 'outline' is prepared (seed formation, ripening).

For medical purposes it is important to know from which stream substances come and how the pharmaceutical process connects with it. What does the separation and changing of individual substances signify? Against this background, how should we regard the conditions into which we bring substances or preserve them when we prepare a remedy?

Grape vine shoot with leaves at various stages of development

Bryophyllum diagremontianum (formerly called *Kalanchoë diagremontiana*). Seedlings are sprouting from the leaf margins.

Differentiation of Substance and Succulence

The processes in a plant that can be traced sensorially and which lead to a more subtle development of its pictorial expression mainly belong to the descending stream in which plant substances arise. The colors, smells, tastes, shapes and composition perceived in the results of the descending stream can give imaginative perception direct indications for the healing effect. But a plant that is taking shape can also retain the capacities of the ascending stream, i.e. of the meristem. This gives rise to various kinds of swelling growth, rather undifferentiated externally, or to succulence. The ways in which vegetative productivity presents itself are closely connected with this. (We will not go into sexual reproduction here). This can be regarded as a growth that becomes separated like a bud or seed and constitutes an independent whole. Something like this happens artificially when a piece of tissue is separated from a plant and another grown from it. This is particularly easy with succulents. The separation is mostly a function of the descending stream that makes organ formation possible so that the higher forms of expression of the organism arise, for example, leaves and flowers. The two streams can be separated only in principle, as obviously the acquisition of form also belongs to the building up aspect. The drop-forming tendency of water corresponds to the process of succulence in plants. Apart from the meristem, all plants tend to be succulent in the upper parts of growing shoots, which are kept upright almost exclusively by the damming up of the flowing sap (turgor) until lignification of the stem takes over. The tendency for glassy swellings in the shoot is sometimes retained, as in Balsaminaceae and *Geranium robertianum*.

The development of succulence, a locking away of the living developmental potential of substance, is usually connected with high reproductivity of a plant and its ability to become physiologically independent of the light-dark rhythm. Succulents can close their stomata during the day and, independently

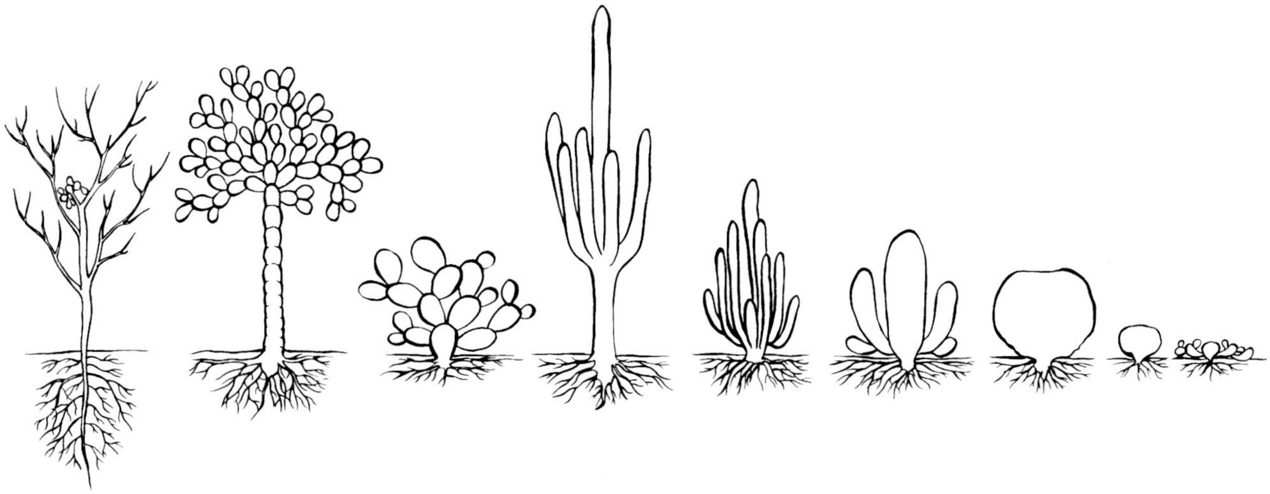

Stages of increasing succulence in cacti from *Pereskia* to *Opuntia* to *Cereus* and the spherical forms

of light effects, process the carbonic acid that they absorbed in the night.
One can compile a whole series of increasing succulence, because it occurs in almost all plant families. In a training course research project, Matthias Uhlig has done this for cacti and has shown the connection with the respective courses of development (see *Elemente der Naturwissenschaft* **51** (2), 1989, 14–24). It goes as follows:

- swelling shoot growth;
- simple leaves do not differentiate further;
- leaves either do not develop or only become thorns;
- whole side-shoots are often left unformed, i.e. retained in the primordium;
- some flowers are taken into the shoot (new leaves appear);
- like water lilies, cactus flowers dissolve in a multitude of petals and other flower organs and during blossoming, cannot retain water.

a b c d

Stages of increasing succulence in cacti at different stages of development
(a) bushy cactus (*Pereskia*), (b) fig cactus (*Opuntia*),
(c) columnar cactus (*Cereus*), (d) ball cactus (*Coryphantha*)

Matthias Uhlig: 'Aspekte der Bildungsweise und Substanzqualitäten sukkulenter Pflanzen unter besonderer Berücksichtigung der Kakteen', *Elemente der Naturwissenschaft* **51** (2) 1989, 14–24

Forms of germinating seedlings

Shoot forms of one-year-old plants

a b c d

a b c d

Schematic nodes bearing leaves and eyes; below for comparison an ordinary dicotyledonous plant. In the axis of the bract, instead of an eye an areole (contracted side-shoot) develops, whose leaves turn into thorns. The bract shrinks and its function merges with the shoot.

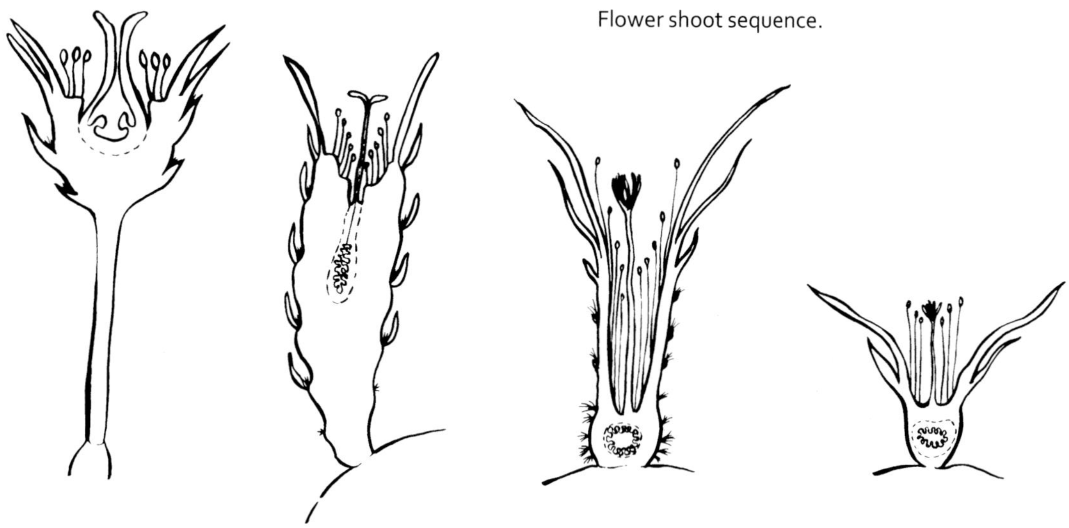

a b c d

Flower shoot sequence.

Petals from *Thelocactus schwarzii* (above) and from a water lily *Nymphaea alba* L. (below). Cactus flowers do not have a fixed number of petals. There is no strong distinction between bracts and petals in either cacti or water lilies.

As we saw with leaves, here we can see that what develops outwardly in a plant's normal development and what can be divided into several developmental steps, is increasingly held back in succulents.
By keeping the creation of substance more fluid, the swelling of the rising stream becomes connected with the process of internalization, which comes to expression in the formation of fruits and seeds.
When succulence is quite advanced, we find a situation that is similar to that of a seed, which owing to its separateness, we actually cannot understand through itself. At least, differentiated percepts of shape – often of colors too – retreat into the background. In such cases, related species categorized into series can provide a visual aid when looking for healing effects. Then we have to inwardly recapitulate the changes in the resulting qualities, actually experience them. What solidifies and takes on form in a special way in related species is mostly kept in fluid form in succulents. A comparison of two leaf series and whole plants (pp. 71–73) shows, in a similar way to the development of succulence, how in the cultivar the formative tendencies of the initial and final leaves seem from the leaf series of the wild relative as if they have been pushed together, and the middle, indented part of the series removed to be replaced by a luxuriant development of unindented leaves. Something similar can be seen in *Cichorium* and other cultivated plants.
In essence, the spreading out of a leaf into the environment is just like a massive, swelling growth
The spreading out of a leaf towards the ground is like an anticipation of fruit formation, with loss of reproductive power and depositing of nutritive substances that can later be drawn on.

▷

All the leaves in the main shoot of cultivated lettuce from cotyledon to flower, *Lactuca sativa* L., variety Reskia (see top p. 71)
Below it a leaf series from prickly lettuce, *Lactuca serriola* L., the wild form from which the cultivated lettuce was derived

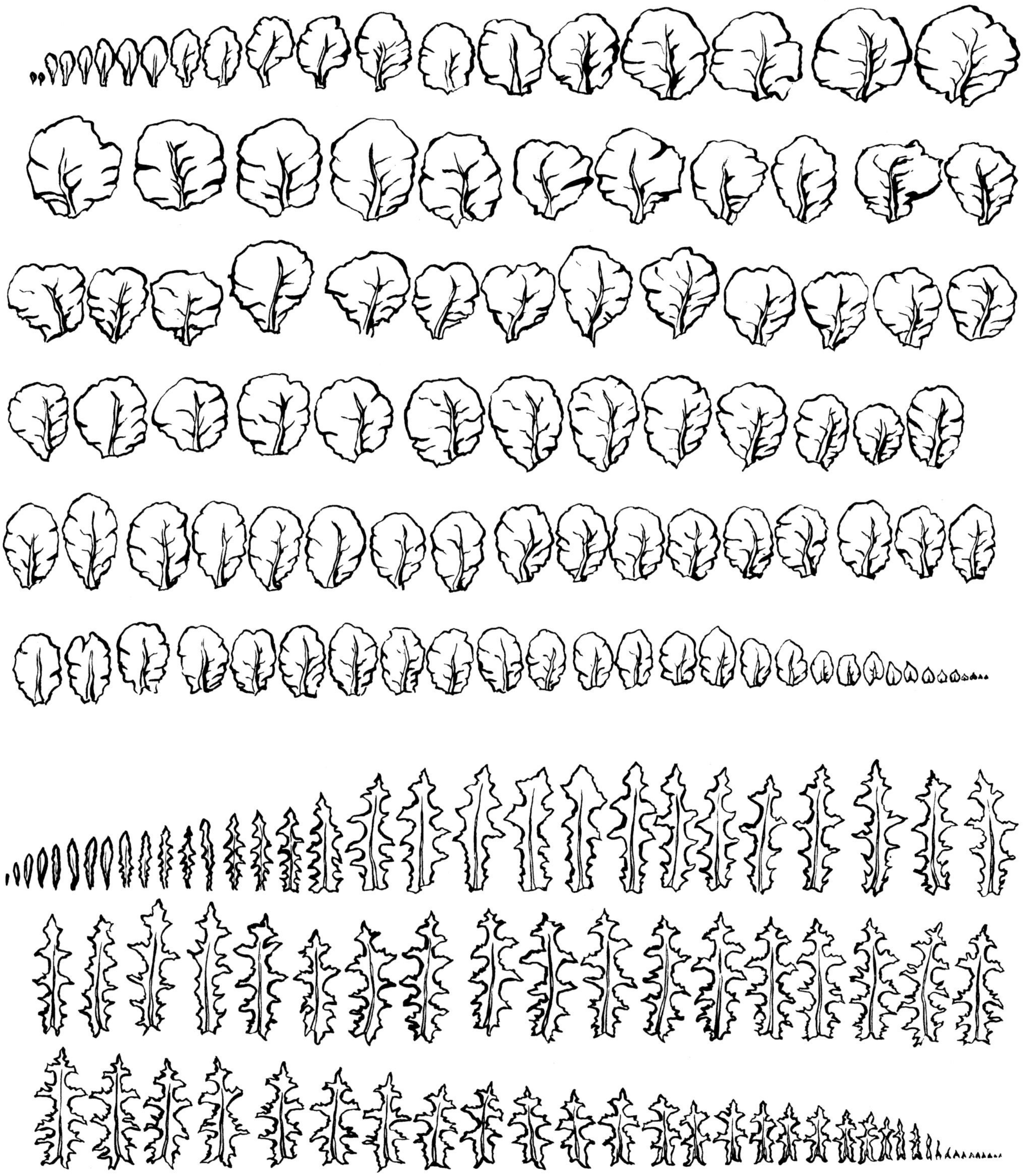

Development of cultivated lettuce: five leaves selected from the beginning, middle and end of the series; early widening of the blade and rounding of the base

Ulrike Behrendt: 'Ein Vergleich der Wild und Kulturform des Salates', *Elemente der Naturwissenschaft* **39** (2) 1983, 15–22

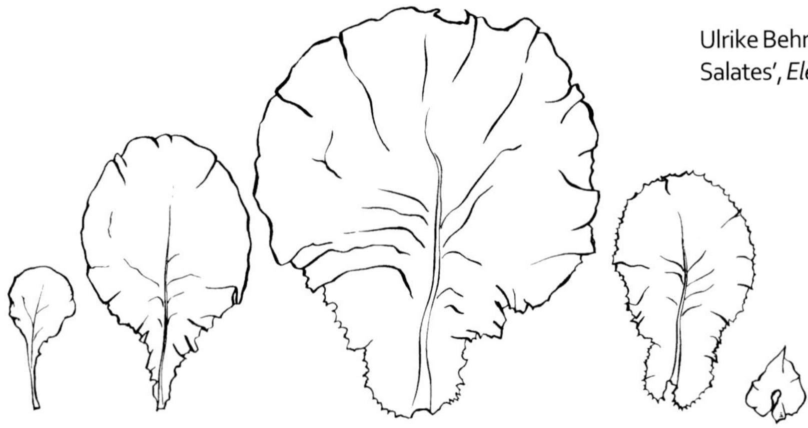

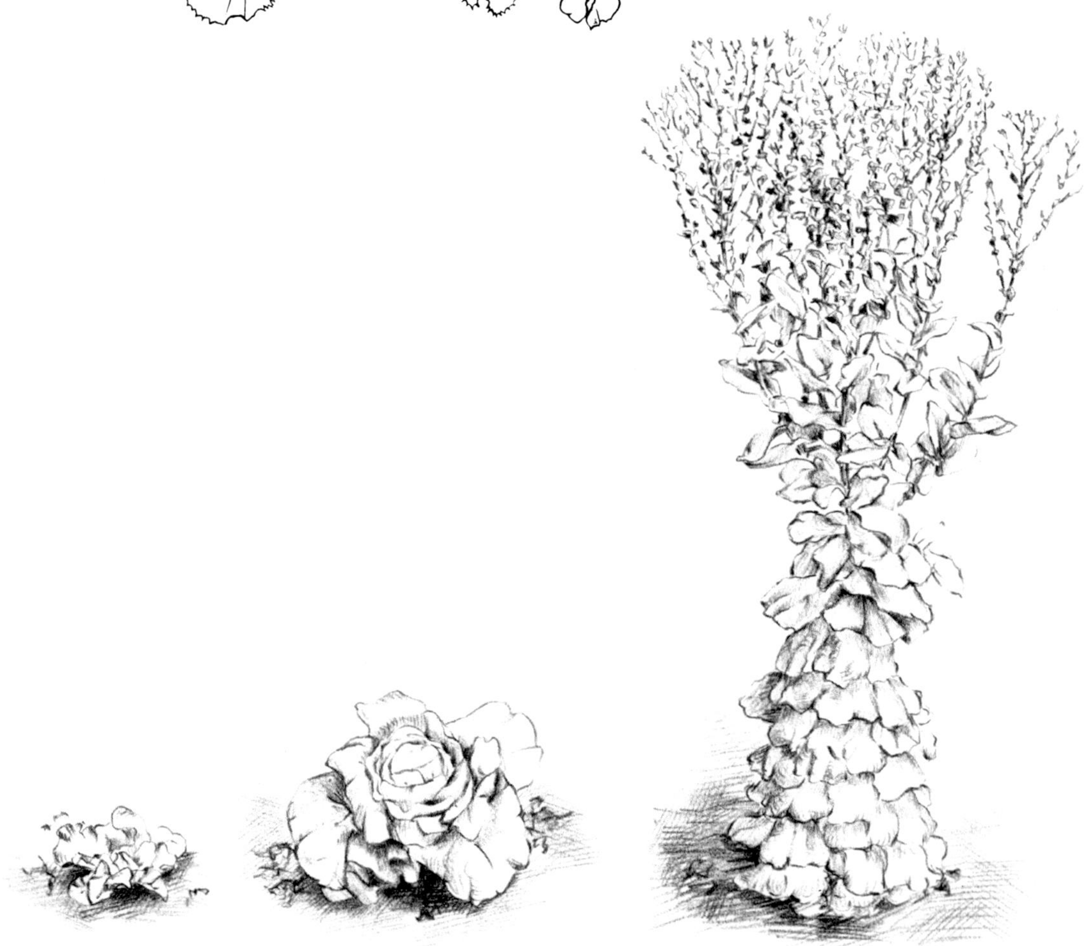

Development of prickly lettuce: five leaves selected from the beginning, middle and end of the series; strong indentation and changing of forms

Inspirational Cognition

A plant that moves us because of its particular beauty is like a being who has something special to tell us. But how do we learn to understand this language? We also know and use in everyday life the capacities that can be attained on our path of schooling. But they are not well developed and what we perceive with them remains diffuse and makes us uncertain. For instance we have the ability *to experience something* from natural phenomena. We experience in them a being's real content (which, however, we immediately push aside with a reductionist description). We become aware of a language to which we are attuned through earlier, personal experiences. But then how exactly is the other being perceived in it? First we react with sympathy or antipathy. The long path of faithfully reproducing plants using imaginative perception where we repeatedly return to the sense percept, teaches us to suppress our initial sympathetic or antipathetic urges, in order to develop the experiencing process in our own activity in the context of the phenomenon. Through this training, something like an environment for the first impression that arose in the sensorial encounter with the plant gradually forms in our soul. This 'environment' is like a linguistic context that resonates in all the pictures and transformations of pictures. As described above, with the soul we grasp the particular spiritual aspect, the character, which we learn to experience ever more concretely through its effect on the soul. Through meditatively building up symbolic pictures from such experiences, we develop the ability to hold mobile mental images in an experience-saturated overall perception, to keep them alive for a while and to deepen them. The overall perception of the plant that we already had an inkling of in the 'first impression', at the beginning of this cognitional process, changes into an organ through which every detail contributes to an experience of a common gesture. This is the clarification of an experience. Everyone has had this experience: it resonates in the soul when we speak of 'yarrow' or 'chamomile' for example. It can be called inspirational cognition.

There are creatures that we experience by direct inspiration (i.e. they speak through gesture), namely animals. They hold a different position from that of plants or minerals in the context of natural life. In animals too we find a pictorial language that speaks to the soul. When we look at the forms and behaviours of crickets, dogs, or frogs, they are not only soul pictures like plants, but also we grasp by direct inspiration something of what lives in their souls. Of course, we should guard against turning these experiences into falsehoods of anthropomorphism. We can avoid this if we do not speak of feelings that an animal might have. If instead we remain wholly in what is observable, in that we immerse ourselves imaginatively in it via the outer form of its organs, and experience the movements ourselves, then we become aware of a goal towards which the animal is striving and we experience what the animal lives in. It is living entirely in what it *signifies* through its movements. In this sense we can also say that an animal lives its significance. But all observations of these processes show that it does not recognize its significance as such. That is, in contrast to the human being, it has no experience of being connected with a self-consciousness.

My feeling life or that of someone else can only be understood and characterized through its relation to the phenomenal world.

The difference between plant and animal lies in the fact that in dealings with a plant an experience arises of how it is formed and how it manifests itself. In this is revealed its specific character as a picture. The picture is like a gesture that, seen spiritually, is produced just as much by the environment. Since the particularity of a plant reveals pictorially its particular environment, we relate this experience to an invisible soul environment that is active in the picture as 'light', 'soil', and so on. The astral effect conceived this way lives in the environment and becomes a picture of the effect of the materialized plant.

See Jochen Bockemühl in *Elemente der Naturwissenschaft*:

- 'Die Bewegung des Tieres als Ausdruck seiner Innerlichkeit', **27** (2) 1977, 32–45;
- 'Vom Orientierungsverhalten des Tieres und seiner inneren Beziehung zu Erde und Kosmos', **28** (1) 1978, 1–12;
- 'Die Orientierungsfähigkeit der Tiere als Wirkung des Zeitleibes (Ätherleibes)', **30** (1) 1979, 22–5

As human beings we feel united with the world and its temporal changes. We find the likeness of the world's archetypal image in our own being. This likeness is no sensory or pedantic imitation of the archetype. What the great world reveals in its temporal flow corresponds to a pendulum swing in our being that does not move in the element of time. Our sensory and perceptual being, we feel, corresponds much more to the nature of summer, woven through with light and warmth. During winter's existence, we sense ourselves much more grounded in ourselves and living in our own thought and will worlds.

Rudolf Steiner, Preface to the first edition of the 'Calendar of the Soul'. 'Anthroposophischer Seelenkalender' (1912/13) in *Wahrspruchworte*, Dornach, 1991 (GA 40); trans: *Calendar 1912/13* (Great Barrington: Steiner Books, 2003)

The Course of the Year as an Aid to Developing a Feeling Interrelationship with the World

When we speak of spring, summer, autumn and winter we mean pictures of warmth, light and weather conditions in connection with the successive flourishing and withering of vegetation. But we also mean the moods with which we live in these pictures. These pictorial moods determine each other. We cannot speak of summer without knowing something about winter, or of spring without knowing the other seasons. This belonging together constitutes the unity of the course of the year, just as we learn of a plant's totality through its successive developmental stages. The individual phenomenal forms of a plant suggest in each case the various ways of approaching it. Looking at a seed demands a mobile, non-spatial perception of a plant's formative capacities. When we look at a flower form we see the result of these formative capacities that arose through interaction with the effects of the plant's spatial environment.

This results in a corresponding spiritual extension of the visible aspects. The varying relationship of a plant to its environment becomes expressed in our soul attitude. In the case of a seed, when we think of the plant we feel that we are in its non-spatial, purely spiritual surroundings and in the case of a particular form, in the realm of forces that are spatially active, either centrally or peripherally. Thus we experience the specific spiritual aspect of a plant species, its cosmic-astral nature, from these two sides, as the determining of its sensorial appearance (i.e. substances), in form, color, smell, taste, etc., and as a purely spiritual tendency of formative possibilities of germinative forms (seeds, eyes and meristems). A plant lives in the breathing transition from one direction to the other.

Accordingly our soul makes pictures of plants that result from its immersion in sense phenomena and from what arises in contemplation from within. In this way it participates in the life of plants.

Our soul lives in this fluctuation. Only feeling can find the spiritual direction in which the pictures and signs are pointing and indicate in what kind of light it is to be viewed. The more shadowy *imaginative experience* becomes inward illumination through a *sense for the language of gestures* that becomes manifest in it. However, no knowledge of the connections between world phenomena and the human being arises as long as things remain, as in ordinary cognition, at the level of merely pondering these processes together.

In his *Calendar of the Soul*, Rudolf Steiner gives us a way to develop this soul capacity of inspirational perception through consciously experiencing the course of the year. Successive pictures of soul moods that change stepwise against a background of the whole year, which lives somewhat dreamily in us, lead to a path of feeling self-knowledge.

Rudolf Steiner contrasts 'our sensory and perceptual being' with being 'grounded in ourselves and living in our own thought and will worlds'; the 'nature of summer, woven through with light and warmth' with the 'winter's existence'.

The weekly verses, arranged according to the season, help in extending our feeling of the differences as we contemplate these relationships. With such empathy we can look at a particular place – especially at what is happening in the plants there – as representative of the changing seasons of the earth and experience more concretely the process as a whole, as part of a breathing process.

The path we take with our feelings leads us to the *soul aspect of the earth*. What happens in the plant world, what drives them out of the earth, is from this viewpoint experienced as soul force that enables the species-specific nature of a plant to become physical. This path can be experienced in pictures, signs and gestures. In general the matter can be described roughly as follows:

Thus, the rhythm of inner and outer becomes for us what nature in its temporal alternation presents as summer and winter. A great mystery of existence can rise if we bring our timeless rhythms of perception and thought into correspondence with nature's temporal rhythms. If we do so, the year becomes the archetype of human soul activity and thus becomes a fruitful source of true self-knowledge.

Rudolf Steiner, Preface to the first edition of the 'Calendar of the Soul'. 'Anthroposophischer Seelenkalender' (1912/13) in *Wahrspruchworte*, Dornach, 1991 (GA 40); trans: *Calendar 1912/13* (Great Barrington: Steiner Books, 2003)

In *Winter* the earth keeps its life entirely within itself. Plants have become withdrawn from the effects of their spatial environment. In relation to the earth, seeds, buds and roots seem to have been inhaled. The effects of the sun from the past year are held together spatially within and have become transformed into new germinating power. This is a result of a meditation on the plant phenomena that were experienced in the past year.

V Substance Formation and its Perception at Various Levels

In *Spring* the earth begins to exhale. We experience how it drives plants out, each in its own way. They are met with sunlight and development begins. A change in consciousness takes place as we tune into this process. Our attention is directed to processes in the plant kingdom and away from wintry, closed off, contemplative rest. An expectant mood predominates. At first, we still think it possible to follow the development of each emerging shoot, but then they become rampant.

By *Summer* the earth has exhaled completely. The soul aspect ('space forming' aspect) of the earth is saturated with the effects of the cosmos mediated by the sun. Plant substances are formed entirely out of these forces. The greens have darkened. Our feelings are drawn into the manifold characteristic phenomena, shapes and relationships to the surroundings. Everything lies on the surface and engages our thinking perception.

With the coming of *Autumn* colors, ripe fruits and withering, plant formation begins to pull back from direct sun effects. Woody things, fruits and new buds become separated in various ways, and past forms continue for a while until they become dissolved in the general humus that is the foundation for new life. We can work with things that have become durable, for example wood for building, or even medicaments. The interiorization of what has been experienced, and its transformation into the life of the mind, prepares for the situation in winter.

There are various levels of the earth's breathing process experienced through plants; the general change of forms and colors in the growing and withering of vegetation and the relatively faster appearing and disappearing of the colors and forms of flowers. These often emerge quite suddenly out of the general green of a meadow or garden and then sink back into it after a shorter or longer time. The one gives a change in the basic mood and, against this as background, the other is like changing facial expressions which tell us something.

Such considerations help form an organ of perception that can be turned towards particular plants. Each plant species has a copy of the seasons in it and thus has a particular relation to this overall process. This is of *significance* for the substance that in future we want to use medicinally.

Jochen Bockemühl, *Erwachen an der Landschaft*, (Dornach: Allgemeine Anthroposophische Gesellschaft, 1992) Ch. IX; trans: *Awakening to Landscape*, (Dornach: Natural Science Section, 1992)

All of these processes would lead to relatively general pictures of plants if something in the nature of a gesture, 'preformed' in the germ of the plant, did not come to meet it. This is, on the one hand, the division of a plant into its organs as demanded by its connection with the course of the year and, on the other hand, the sphere of the laws of form whose framework of relationships underlies the system of plant families for example.

The fixed stars present a picture of how soul and spatial aspects are related to each other via the sun. There is a spatial orientation for knowledge of the totality (inspirational). Single stars are like points on a map. The meaning of points on the map becomes evident only if, using the map as a guide, we actually move in space to the corresponding places and there take in whole vistas. For our perception, the stars seem to be on a surface. Through their interrelationships with each other and the earth, they enable us to orient ourselves in space (navigation). Meditation on their relationships also leads us to an experience of spiritual regions, from which we can enter purely soul-spiritual realms. Plants can refer us to these spiritual regions in a pictorial way (as gestures).

V Substance Formation and its Perception at Various Levels

While it is true that dried, shredded and packaged plants or processed medicaments have been lifted out of their original context and we can make free use of them, we actually know full well and assume in practice that plant substances always arise at a particular place and time in a natural context.

Transformation of Earthly Substances into Living Substance and into Soul Substance

We perceive earth substances through our senses and assess them through their relation to other substances. When Rudolf Steiner speaks (1925) about the streaming of terrestrial substantiality into the physical, etheric, and astral bodies, he goes beyond what is merely sensorial, objective. We have to understand terrestrial substantiality then in such a way that palpable material takes on the inner nature (essence) of one of the bodies. When a mineral assumes the etheric nature of a plant this means that certain *capacities of live plant substance* are imaginatively revealed through the way that shape, color, smell, composition or taste manifest and change. Plant substance only exists through the acquisition and use of its capacities. Substances that we know belong to the physical body of a plant are either on the way to it or are kept away from it and, in the separated condition, no longer in a living context. Echoes of the latter can remain, however. This applies to conceptual separation when we refer to *constituent substances*.

Rudolf Steiner and Ita Wegman, *Grundlegendes für eine Erweiterung der Heilkunst nach geisteswissenschaftlichen Erkenntnissen* (GA 27), Dornach, 5th edition 1977; trans: *Fundamentals of Therapy* (London: Rudolf Steiner Press, 1996)

When earthly or living substance in an animal takes on the inner nature of its astral body, inspirational expressions of the soul (see p. 75) are experienced in the interaction between organ shaping and behaviour, i.e. become sense perceptible, and they also speak inspirationally which means their significance is lived. This is especially clear where, for example, the animal's behaviour repeats itself, as in the construction of a spider's web.

The behavioural change in the spider under the influence of a drug shows in the changed and chaotic

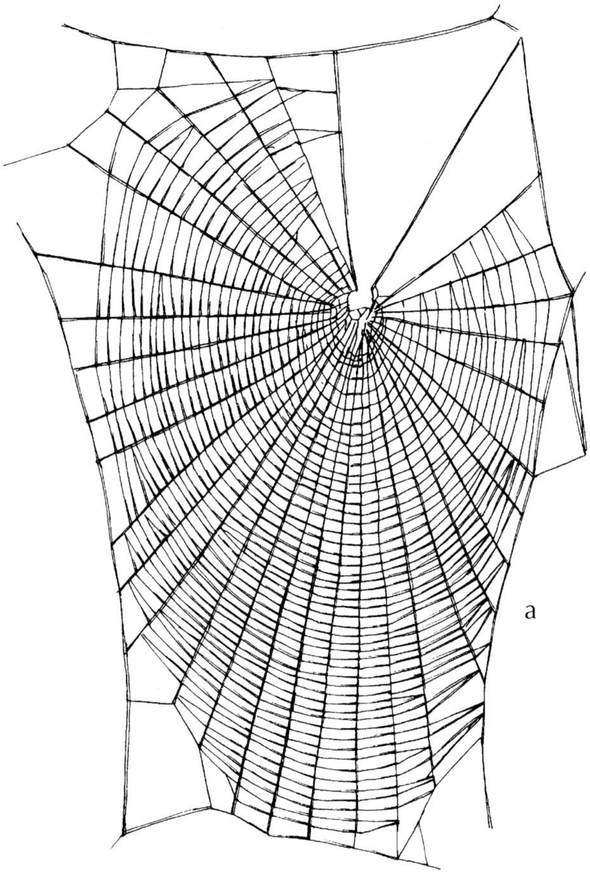

a

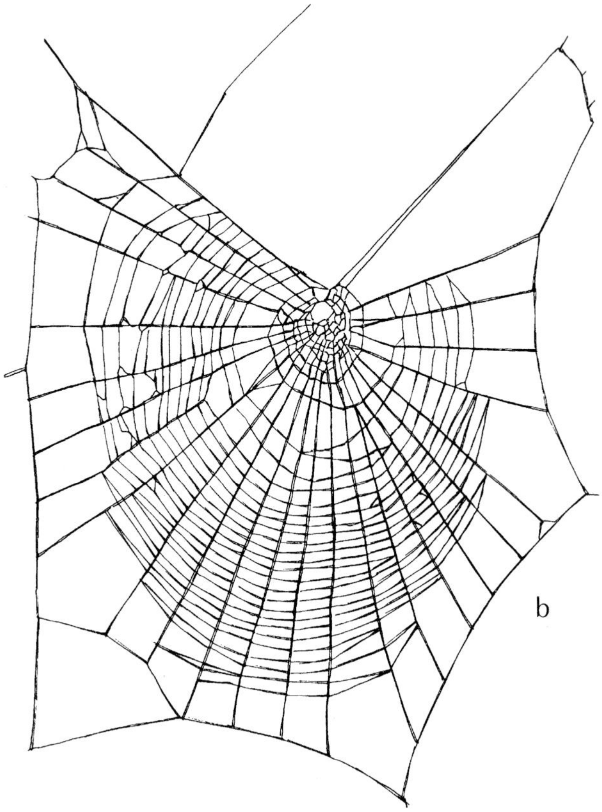

b

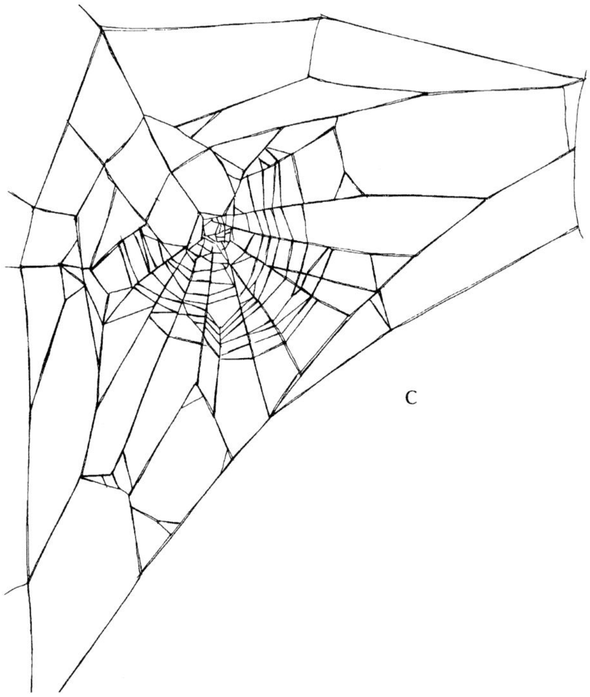

c

a) Normal spider's web
b) web of a spider under the influence of mescaline
c) web of a spider under the influence of caffeine

Drawings from Witt/Reed/Peakall: *A Spider's Web*, Springer 1968.

forms of its web. A plant substance comes to pictorial expression in each case. An observer can have a soul experience from this. Thus plant formation reaches the threshold of the astral world, and it appears to be shaped by it. It produces (astral) *effects* on the soul, without itself (as with animals) developing an inner life, not even when it creates flowers that look like animals. We can experience the spiritual quality that has become (materially) a picture in the plant; it makes a particular aspect of our soul resound. In this sense, soul and astral aspects are always experienced outwardly as form and inwardly as a particular world of experience.

We experience the existence of ensouled substances in us as feelings arise (are released) in relation to world phenomena. Our ability to let consciousness arise through a detachment from the sensorial mode of appearance is comparable with the autumnal process, for the degradation of a plant's outer form is accompanied by an inner enlivening in the formation of seeds and buds.

Archangel's gesture in *The Annunciation with Lilies (1480).*

From a painting by Carlo Crivelli 1457–1493 .

Intuitive Cognition

The last step on the cognitional path leads to experiences that mostly escape our consciousness. But we do have elementary experiences of this kind that we rely on in daily life.

In *The Philosophy of Freedom* (1894), Rudolf Steiner points out how thoughts arise in consciousness, and he calls this kind of experience intuition. If we think about such a process we notice that the moment we really grasp a thought, thinking, feeling and willing coincide in our consciousness in such a way that there is nothing longer anything 'left over'. We become one with the thought. But as a thought arises in our consciousness in this way we usually overlook the fact that our self has become one with it, that we think *in* the thought as we produce it.

Rudolf Steiner, *Die Philosophie der Freiheit*, (GA 4) Dornach, 15th edition, 1987; trans: *The Philosophy of Freedom*, (London: Rudolf Steiner Press, 1964)

We experience such moments of identification each time that an outer percept approaches us like a question and becomes united with something that

Rudolf Steiner describes the transition from imaginative to inspirational consciousness in connection with indications for a path to a rational medicine:
'Take the relation of the environment to our own inner life. We can describe this in such a way that we say: for ordinary consciousness the material world is out there. If we look at the human being objectively we say: when a man looks through his eyes into this material world and perceives other things through his ears, then the material things and facts are outside, and his ideal contents, his soul contents are in his soul-being in thinking, feeling and willing. When a person perceives what is material they bear this outer material world in their soul pictorially, as a copy, as delicate and light as their soul. The moment we learn to grasp the spiritual world around us in our empty consciousness something new arises for our inner life.

Let us assume at this point that for inspired consciousness I were to see this material world permeated by the spirit world. Now what is perceived spiritually outside does not arise within a person pictorially, but rather we now learn to recognise the spiritual outside in the way it is reflected within the person. It is reflected as their physical organs, as lungs, liver, heart, kidneys etc. and as everything else that is at first material within them. It is a complete inversion or reciprocity. Whereas the material world is reflected in us in a soul-spiritual way for ordinary consciousness, the spiritual world is reflected in us through our organs. We learn to know ourselves inwardly as physical people when we become aware of the spiritual world around us. Until then nothing will be understood of the physical human being. Until then we will get to know the heart, lungs and liver outwardly through anatomy, but not their connection with the outer world. Through anatomy and physiology we get to know the heart, lungs and liver as if we were getting to know that a person has all kinds of ideas within them and yet we do not realise that these inner pictures refer to the outer world. We do not realise that these organs are related to the outer spiritual world.

we produce in our thinking. This is called a *feeling of certainty*. We feel that the intention that lies in the thinking act coincides with what we notice in the other.
Experiences of certainty occur at various levels of cognition. They happen most easily in the experience and judgment of what we call 'matter' or 'object'. Here experiences of certainty are based on *ready-made ideas* that coincide with experiences of the limbs. (My idea of a particular stone is real if I can lift it.) Our ordinary self-consciousness is also based on this. What we support ourselves with in cognition changes when we ascend to imagination. It becomes more inward. Ideas become flexible in *imaginative perception*. Self-consciousness is based on *thinking activity* that agrees with sequences of phenomena.
The connection between a particular feeling that arises in inner mobility and outer observation becomes evident in *inspiration*. This support (evidence) arises on 'listening', in receptivity for 'the other', and in becoming aware of a resonance in the soul, i.e. in a selfless, feeling self-consciousness.
In *intuition* the being with which we connect ourselves lives in our will (capacity to love). This process can only be described approximately, for in experiencing the certainty, there is at first only the 'other' that we are certain about. For in most experiences of certainty the identification only resonates dully as the foundation of our self-consciousness. We are then convinced about something without being aware of where we really stand with the conviction. Hence ordinary convictions are only very dull intuitions which are not suitable for responsible action. Responsibility requires us to connect ourselves with something (plant, person, etc.) in such a way that we can act out of its being.
To do this in intuition, an awakening for our relation to this being must also live at the levels of sensorial, imaginative and inspirational experience. However, it is a long journey to these higher forms of intuition, and, let us be clear about this, we always have further to go.

But the beginning of every cognition is already an intuition. In observing a particular plant we already grasp what there is to know. I call this an 'intuitive anticipation', without which we have no guiding star for knowledge. This is more or less the same as what we often call a 'first impression'. Here a momentary overall experience of the other creature arises and its echo can then become a guide to knowledge. Thus something initially lives in every intention with which we have intuitively connected ourselves. Intuition is never something finished, and yet it is always the whole in each case.

The Experience of the Threefold Organization of the Human Being in Relation to Observations from Exercises on Plant Formation

The stages of self-experience mentioned have their expression right into a person's physical shape. For instance, we experience this if, geometrically and pictorially, we transform the structure of the limb-system into that of the skull, and in so doing become aware of how in life we express our soul through these forms. We experience the way in which we are active with our limbs in the outer sensorial, physical world, and how our brain enables us to enter an inner relationship in thinking with the spiritual world. And with our rhythmic system bring ourselves into harmony with the experiences that we had in both directions.

Through these differing approaches, appropriate for each case, what we experience in the kingdoms of nature becomes permeated with our self-experience and thus becomes accessible to self-observation as inwardly experienceable gestures. This helps us to understand the threefoldness of the human organism (nerve-sense system, rhythmic system, and metabolic limb system).

This is the origin of what makes it possible for spiritual science to work in a rational medicine, for example. For only now do we really get to know the human being, become acquainted with the inner nature of the human organism. Previously we could not get to know it at all, just recognise it outwardly'.

Rudolf Steiner, *Damit der Mensch ganz Mensch werde: Die Bedeutung der Anthroposophie im Geistesleben der Gegenwart*, 'Die anthroposophische Forschungsmethode', 10th April 1922 (GA 82)

Experiences towards an Understanding of the Nerve-Sense System

As we showed with plants, in its ascending stream an organic body tears itself loose from the earth organism. Mineral substances are converted into living ones towards the meristem. In the descending stream an organic body is built up and maintained out of its etheric and astral environment in forms specific for the species. The life of the human body is based on the same principle.

Something counteracts this when the person is awake. In our own formative activity with percepts, forms are taken in, and in understanding the way a percept becomes a picture of what is observed, we learn to recognize the nature of life, of the etheric. Here the life of the human organism is suppressed or deadened. Consciousness of something else only arises when our own life recedes.

If we reflect on the way an outer percept turns into inner perception, it leads to an experience of a way of working that Rudolf Steiner describes in the extract shown alongside. The 'soul formations' first appear as a picture in the sense of inner perception. Through thought perception in the sense of the preceding exercises our direct imaginative understanding shows us how we can grasp, as a lively capacity of consciousness, what leads a plant into shape formation. What takes on physical form in a plant becomes inner experience as a spiritual picture in us. But for this to take place the physical life of our organization must partly come to rest.

This plant formation, therefore, is the opposite of what happens in human thinking. Impelled by the sun, a plant absorbs dissolved salts from the dark earth. Mineral is enlivened in such a way that newly formed substances differentiate out through the etheric-astral effects of sunlight. During this process the plant takes on forms via the earth which are bodily shapes and which can be experienced by us but are not in themselves experiences.

'In conscious thinking, thoughts come free of the physical configuration process and as soul configurations become living experiences.'

Rudolf Steiner and Ita Wegman, *Grundlegendes für eine Erweiterung der Heilkunst nach geisteswissenschaftlichen Erkenntnissen* (GA 27), Dornach, 5th edition 1977, p. 17; trans: *Fundamentals of Therapy* (London: Rudolf Steiner Press, 1996) Ch. 1, pp. 10–11

Thus, experiences gained through thinking that has been strengthened by exercises create imaginative insight into the workings of our nerve-sense system, which is always on the verge of dying, so that it can serve as an organic mediator of consciousness.

Experiences towards an understanding of the rhythmic system

The formation of an animal organism shows that it is organized for movement in a certain way and that it is only because of this that it is connected with the spatial environment. In the shape of an organ as well as in the pattern of an animal's movement that harmonizes with it, this relationship with the environment can become a direct inspirational experience. For instance, a horse's limbs are different from those of a fox, a fly or a snail. A horse pounds along; a fox slinks, touching the ground softly; a fly clings, a snail glides, and so on. That is, they all deal differently with the earth and the other elements. Moving like this and expressing their nature, they are dedicated to the environment to which they are bound with their senses – the horse to the water trough, the fox to the catching of a chicken, and so on.

When we consciously transform ourselves into the particular animal, as it were, (which we usually do anyway unconsciously when we look at it), the particular element-qualities and sensory connections become inspirationally experienceable through the patterns of movement in which an animal lives (and thereby experiences its 'significance'). In our consciousness, the animal perceived seems to disappear into our experience of the signifying gesture of its movement. We experience its soul or astral nature in the movement that is then seen embodied again outwardly as a horse, a cat or a snail.

As we have shown, the way a plant is formed is connected with its cosmic-terrestrial surroundings in the course of the year. 'Moods' of the landscape and gestures of the etheric-astral environment are copied in plants as they shape their leaves and

shoots one after the other and bring life to expression in rhythmical changes in the transition from one to the next.
Reflecting on such 'breathing events' leads to *inspirational* knowledge of the way the *rhythmic system* works in the transition of 'soul formations', in imaginable pictures, in the actual, imageless, soul-corporeal events and experiences and vice versa. We carry out this process continually in our feeling experience of the world. An artist is active in it.

Experiences towards an understanding of the metabolic-limb system

As shown in the previous chapter, every experience of a reality also has an intuitive character. To the extent that human knowledge is accompanied by self-knowledge, intuition lives in it: something is certain because we are united with it right into our body and our will. This is directly the case in the activity of our limbs. The same thing takes place as a percept when we meet someone, and indeed most clearly where the personality can actively reveal itself through actions. The personality has a character (astral body) and an individuality (the I). The extent to which the individuality penetrates its body right down into the metabolic processes becomes intuitively graspable in speech, movement, mime, and flesh color. The I reveals itself through its mastery of the power of will in the degradation of the body's substances. The more freely this takes place the easier it is to feel our way intuitively into other beings. What in a person is an expression of the innermost being (inwardness) becomes, when it is turned round, an outer picture in the shaping of substances. When we look at flower formation, we have the most characteristic part of a plant before us as a physical shape.

The formation of seeds that are like a summing up of the whole plant is prepared within it. If we follow this process from flower to seed formation we get as close as is possible to the *intuitive* aspect of a plant's being. We attain a soul attitude, which we can then apply generally to what is happening in the plant's main axis from roots to seeds.

Plant Formation and Human Organization

By going again through all of the steps that were presented for an understanding of the human organization in relation to a plant we can discover the following:

- The physical process of root activity in relation to the part of a plant above ground reveals imaginatively the etheric activity that mediates between perceiving and thinking as an inverted picture of the nerve-sense processes.
- In grasping imaginatively the way in which plant organs are formed, astral activity begins to speak inspirationally in the step from the metamorphosis of the leaf-organ to flowers. In the processes of the rhythmic system new impulses for action, and ultimately for forms, arise from the experience of the spiritual form taken in by the human being in perception. This means enlivening and dying in differentiated physical forms; then re-enlivening with the new transformative impulse that is gained in reshaping and new shaping.
- In our inspirational study of the transition from the gesture of a plant in flower to formation of the seed, in which this gesture becomes a hidden formative capacity, we approach an intuitive experience of the being of the plant.
- This last principle can be traced vertically downwards again. If, starting from seed formation, we experience the processes of the formation of flower, leaf and root in this way, an initial inner experience arises of the various ways in which people live in their organs as they think, feel and will with their soul.

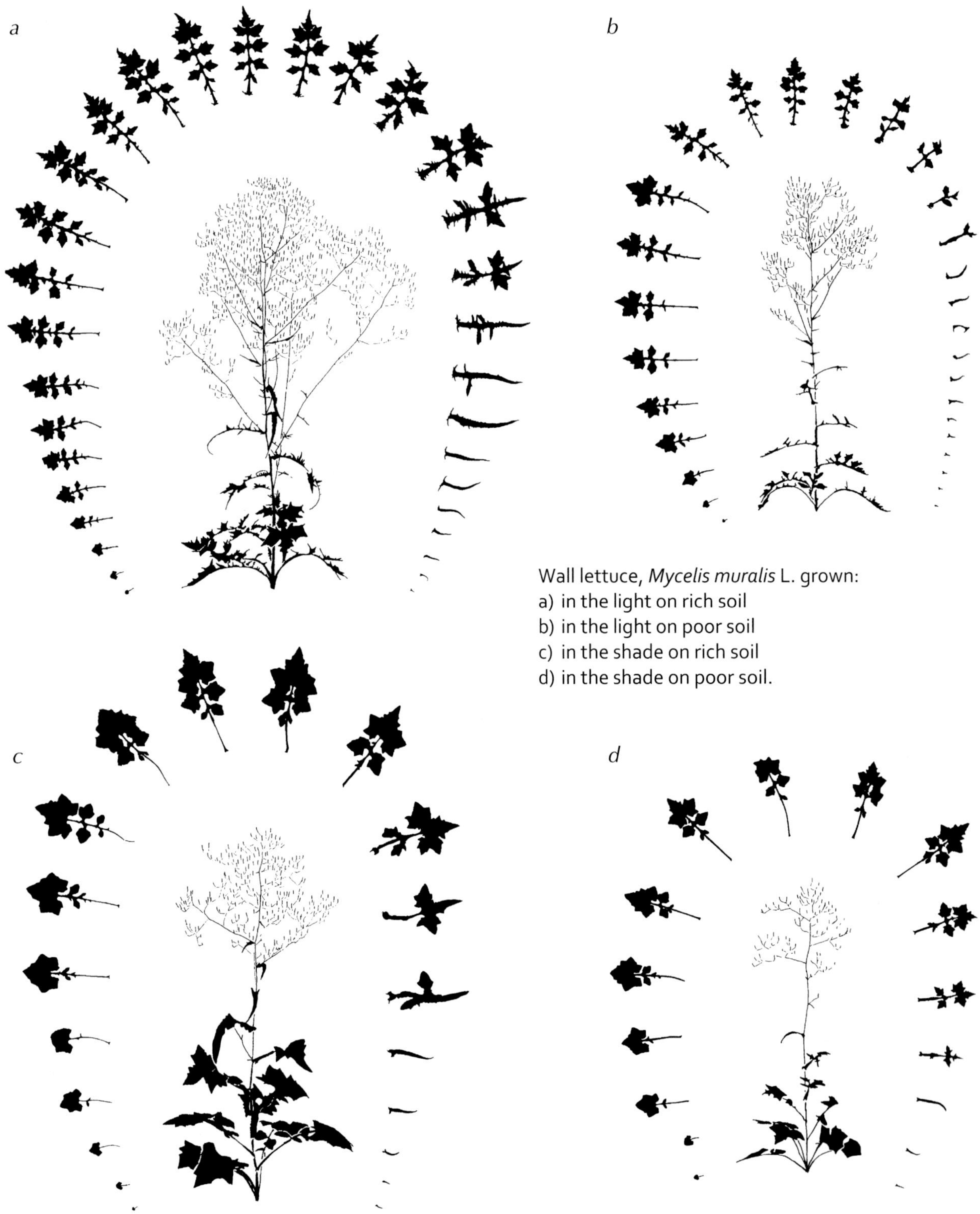

Wall lettuce, *Mycelis muralis* L. grown:
a) in the light on rich soil
b) in the light on poor soil
c) in the shade on rich soil
d) in the shade on poor soil.

Connections with the Environment

On this path of allowing plant formation to become active in our soul, we press on to ever more comprehensive 'environments' of plants.

1. When we observe plants in lighted and shady, wet and dry, fertile and barren places we find out how substances develop under the influence of meteorological conditions. We get a mobile picture of the effects of the elements (light, soil, seasons, etc.). Here we take three things for granted: that plants are different from human beings, that species have particular characteristics and that life processes are differentiated.

2. If we move on to actual becoming and passing away and to experiencing the descending and ascending streams of substances in a plant, we are dealing with a differentiation of life processes under the influence of cosmic-etheric effects which, via light and warmth (sun), make things manifest and which, via earth and water (moon) and other planets, create a readiness to receive them. We become aware of the connection between inner activity and receptivity in our soul. Here we are still presupposing the plant in general and its specific character.

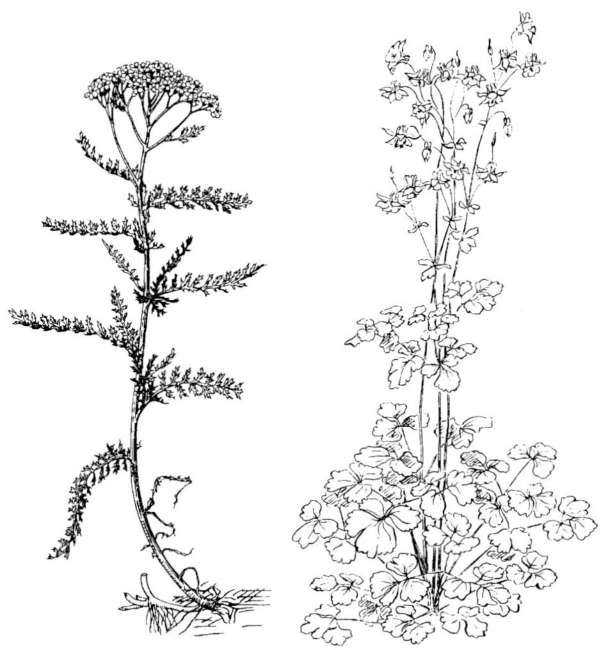

3. Through a conscious experience of the particular expression of a plant species, we arrive at its astral environment. It is the world in which the formative principles have an experienceable relation to each other, and it is against this whole background that the individual formation becomes a gesture. Or, putting it more imaginatively, every plant species becomes a 'star' viewed from which the world acquires a particular coloration. In our own soul this corresponds to a particular way of looking at the world.

4. The whole of the starry sky is the outermost cir-

cumference of the plant world. This is the sphere of the 'archetypal plant' which can never become visible as such. It is only here that the nature of plants becomes one with that of human beings. What takes place in connection with a plant's main axis is our place of spatial reference for this experience.

When we observe how a particular plant lives with its shoot and root, we can see its particular relation to the earth, i.e. whether it spreads out spherically or goes more to the earth's center. On the other hand, the way it creeps along the ground or lifts itself up and develops its leaves, flowers and fruit tells us about its relationship to the environment. Thus a real plant necessarily becomes one-sided through its shape and substances, i.e. individual in an earthly sense. This inwardly sought environment can then harmonize with the mood we can experience at the place in the landscape where it grows. By tracing the 'one-sidedness of plant formation' we notice a connection with the one-sidedness in our own soul.

Nutrition and Healing

People have to build up their bodies and organs with what they take in from the environment. This is the normal process of nutrition in the widest sense, which will be further differentiated in the following. It is a kind of challenge or support for a human organism, with which its life processes are sustained. Healing through a particular substance is basically a special case of nutrition. A malfunctioning human organism has to be given a *specific* challenge to put it on the right track towards helping itself again. Plants are an important source of such substances. That is why the following discussion is restricted to plants at first and only later related to minerals.

Education of the Soul via Perception as the Principle of 'Cosmic Nutrition'

We have already begun to take in something from a plant in everything we see of it. In doing this, we are touching upon the secret of 'cosmic nutrition'. This means that everything that we take in through the physical senses – above all through seeing – stimulates something in our organism that lives on in the soul. With our receptive soul we dive down into percepts that are at first foreign to us. The life processes of the eyes react and thereby stimulate an aspect of our soul. Thus what is at first only a tendency to receptivity gradually gains experiential content.
In this way a sense percept works on our soul body (astral body) via the etheric body and lets the world come to life in our soul, which is enriched and acquires more contours. Then if we judge what we have experienced, with our self-consciousness, we again distance ourselves somewhat from this experience; we spiritually place it, so to speak, in a particular location and give it a form.
For the aspect of cosmic nutrition, we direct our gaze at the processes that occur in our soul *before a judgment* and which shape our soul body. They

This sequence of pictures from Spain shows that we can experience the various moods of early spring until mid-summer almost simultaneously as we go from north to south and from the mountains to the seacoast.

On the pass of Sierra de Guadarrama (1800 m) north of Madrid. Crocuses in flower and small, short-stemmed narcissi beside patches of snow. Middle of April.

can be best observed and described in connection with seeing. What people take in directly in seeing, without changing them physically, are above all a plant's shape and color. Added to this is smell.

At the foot of Sierra de Guadarrama near Miraflores (900 m) north of Madrid. Meadow with compact oaks, asphodel and squill in flower. Middle of April.

Cosmic nutrition through color in our surroundings

Every color quality and color transition allows us to participate in the inner life of outer phenomena in a particular way. Thus the coloration of the landscape and plants leads our soul directly into nature's coming into existence and passing away. We need only think of the feelings that the colors of a spring, summer or autumn landscape arouse in us and which enable us to participate in the seasons.

This participation directly informs the human ethe-

Near Villanueva Pto. de la Pedrizas, pass (780 m) north of Malaga. Various flowers such as brownish red digitalis, peony and dwarf palm amongst dry gravel. End of April

ric body. The landscapes from Spain are intended to indicate how something similar can be experienced at various places at the *same* time. Likewise, we experience the effects of the subtler color differentiations and changes of a plant as its conditions of life. *Feeling capacities* are awakened and developed further as we get involved in these processes.

Summery display of wild flowers on the south coast of Spain with views towards Montes de Málaga. End of April.

Cosmic Nutrition through the Perception of Forms

In the preceding discussion, what referred to the *shapes and ways of formation* of plants is above all *seen*. But what we see we also feel at the same time by touching. We turn to these visual sensations felt by touching with senses that usually only serve the perception of our own bodily processes and conditions, namely, with the senses of our own motion, of balance and of life. Thus we grasp the forms in seeing and we identify ourselves with them in such a way that they are experienced as gestures. The seeing and touching of forms and formative transformations is followed immediately by a kind of 'soul tasting' and 'digestion' but without any intake of a substance.

In this way a soul actively participates in the gestures of sense phenomena. The disposition to receptivity that is germinal within it receives an earthly content and takes on living forms. Slumbering capacities are awakened from within the active human being, i.e. from its center, which increasingly shape the etheric body via the astral body and these then build up the physical body. These are far-reaching effects on human beings. We need only think of how a child's soul becomes organized via seeing (and hearing), i.e. through the *experience of forms* that get 'digested' through imitative activities. This creates specific shaping capacities and extends the child's 'receptivities' or possibilities to take in differentiated, formed examples in life. This leads to the formation of, in consciousness, increasingly differentiated ideas and, in the physical body, the bones.

Downy oak *Quercus pubescens* Willd. ...

Cosmic Nutrition through the Sense of Smell

A plant's formation of substances goes beyond the boundaries of its physical form when it *emanates aromas*. We absorb these via the air as something that is sometimes 'tangible' through our breathing. Therefore, perception in ordinary consciousness is guided by a slight impression that we are dealing with 'matter' taken into our body. This impression of matter confers reality to the bodily process of smelling. The pictorialness of the impression becomes clear in the percept. For instance, 'speaking pictures' are aroused when we smell roses, lavender, or pear blossoms. Here the direct, pictorial impact of the percept is present. But the mental images that are connected with smells are associative like dream pictures, and their content is only weakly connected with what the actual experience is telling us. It is even more important to note how smells put someone in a particular soul mood or frame of mind to do something. The perfume industry discovered that long ago. It is even clearer in animals. For instance, we see that a dog lives almost entirely in the element of particular smells that affect its behavior.

... and white willow *Salix alba* L. Notice the very different gestures of the two trees.

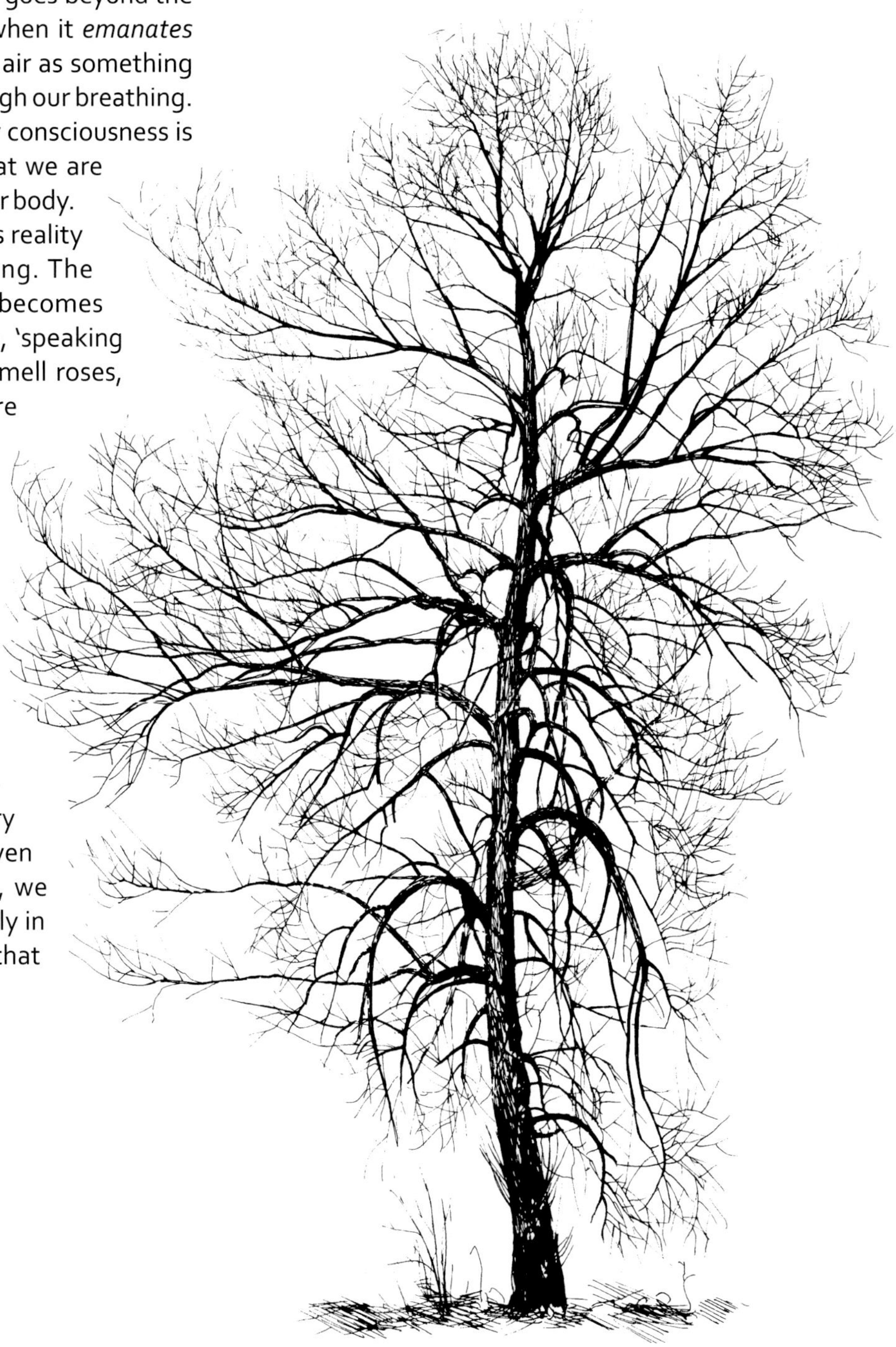

The human soul, especially of a young person, is still very pliable in this respect. What lives in it is not yet entirely fixed and is completely connected with the body. As people get older their soul and mental imaging consciousness becomes somewhat lifted out of the direct connection between smelling and bodily needs.

One first has to learn to understand the real connection with substance by practice in order to keep scent percepts in a factual connection with visual, tactile and taste percepts. Otherwise the connection between the life of the whole person and the world is lost in impulses that arise instinctively through smell. Then they work destructively on the soul's relation to the physical and etheric bodies. We can more generally summarize the effects of subtler cosmic substances through shape, color and smell under the concept of cosmic nutrition, to the extent that they develop the soul as a whole.

The Greek word νουζ (spirit) for 'smelling' is believed to be connected with 'Snottar' (OHG) – wise man, and the stem of snuffling – (s)no(f)os.

Cosmic nutrition via seeing

'seeing' in the sense of a pictorial comprehension	*from the separateness*	*forms and modes of formation*	*grasped through seeing while feeling by touching; made our own as shape and expression with the lower senses; creates capacity to have ideas and specific shaping ability in the life-body and in thinking*
	directly	*colors and transformations of colors*	*experience dives directly down into the imaginative world of colors; participation in life processes including specific soul moods; capacities for feeling developed*
	association with what is visible	*smell*	*hint of materiality; connects soul with body; puts us in particular soul moods; arouses instincts, desires or will urges that want association with visible things; the process becomes more human the more the real spiritual connection with the substance is sought in it through the mediation of the senses*

The human body becomes an expression of its soul, its being, via the soul processes of tasting and digestion that are connected with the cosmic aspect absorbed from the surroundings through seeing. Sense impressions that are out of context can have a destructive effect, especially in children, if their souls cannot preserve the unity with context from within. (Television is an example of this. It works with pictorial impressions that are connected neither with each other nor with the corporeal life of the people who view them.) Specific environmental effects can also be used therapeutically: for example, color therapy; 'gestalt therapy' via movements; landscape composition via immersion in a landscape; therapy via the sense of smell (sea or mountain air) etc.

Cosmic nutrition means a build-up of the soul body through the vividness of the sense world out of the germinal spiritual possibilities of the soul. Capacities arise that shape the body in such a way that the soul can live freely in it.

Physical nutrition, Digestion

Thus, cosmic nutrition shapes a man from within, for it works directly on his soul via his senses. This enables him to develop his organs so that his soul can live in them. Complementing this is physical nutrition and the toxic (destructive) or healing effect of substances that are absorbed via the metabolism. This substance intake proceeds from solids or fluids and leads initially to the degradative dissolution of substances. If only dissolution was at work unopposed by the I and astral body, the physical body would melt into formlessness. Indeed, this dissolving tendency is an expression of the fact that the human organism brings something to meet the substances that come from outside. The stimulus occurs via perception. It begins with touching and tasting and continues in physical digestion.

Physical nutrition via taste, digestion and development of the physical body of forces

Taste: *Entrance of the astral body and I into an interaction with the ingested substances; special affinities are addressed by each organ*

Digestion: *Life processes mediate between inner and outer; the inner must be present, i.e. the soul must have its independence working from within outward, but each time it must give this up to a certain extent and regain it in rhythmic alter-nation;*

Conversion into 'self': into substances of the physical, etheric, and astral bodies and of self-consciousness, through which a man feels that he is in a different relation to the world in each case and becomes active in it.

Tasting as a Transition to the Digestion of Physical Substances

When we taste food that is at first *palpable* we feel closest to the action of substances on our body. Every tasting is sensed as a direct interaction with the body in a particular direction, for instance, in an astringent, anaesthetic or sialagogic way. In tasting, the astral body, which develops itself through cosmic nutrition, is drawn to the physical body via life processes. So taste and smell work in opposite ways. Whereas smell puts us into particular soul moods, i.e. creates a mood of expectancy for food, tasting stimulates the astral body to particular activities from the corporeal side. *Conscious* tasting is the beginning of perception of continuously changing material effects at the transition point to the life processes of the body. It stimulates the astral body and I in making the physical body interact with these processes.
Stimulated by substances, taste gives the astral body directions for this interaction. Ordinary food tastes relatively neutral. Food that tastes good indicates that there is harmony between the need to consume it and a pleasant interaction with it. The body's life sense perceives this. Stronger challenges or directions for the organism come with the use of spices on food. This is related to the effect of a medicament. Tasting continues within the human organism, and undergoes stepwise changes that become increasingly difficult to experience consciously.

Digestion

Tasting is only the beginning of digestion. This is followed by life processes that mediate the passage of substances into the human organism. We will go into this more in Chapter VII, page 122 'Life Processes Connect a Person's Soul and Spirit with the World around'.
The life processes begin a stepwise transformation of

substances (see next chapter) which help the soul and spirit develop both etheric-physical organs and above all physical strength. Thus the following picture arises for physical nutrition (digestion) and complements cosmic nutrition via seeing.

The Stepwise Intake of a Physical Substance into the Human Organism

Rudolf Steiner describes how earth substances flow into a human body as far as the I organization. Here the question arises of how we can observe the course of assimilation of a substance in the human being, i.e. right into the I organization. We should pay particular attention to the stage in the process that begins with the intake of a substance where we are induced to change our way of looking at things by the nature of the percepts. This leads to various kinds of 'tasting' and the discovery of substances at various levels of reality in our organization. Through the path of self-experience and reflection upon it, we need to free ourselves from the usual conception of substance, as we tried to do in the preceding chapter. Each stage has its own certainty that I must experience *for myself* (see 'Intuitive Cognition' in Chapter V, page 85).
Then we need to look for the connection between these experiences by becoming aware of the step taken from one stage to another, i.e. of the changes of direction in our attention that have to be made from one experience to the next. This is difficult because it has become a habit in science to consider only what is analysable as materially real. If we find correlations between these analysable substances and life processes or psychic phenomena then they are interpreted as effects of *these* substances.
However, if we were to really think the correlative relationships we would soon experience the boundaries we come up against with such a cognitional attitude. Thus, from the way the concepts of substances are formed, we cannot think of them as *cause*, i.e. we can never derive the corresponding

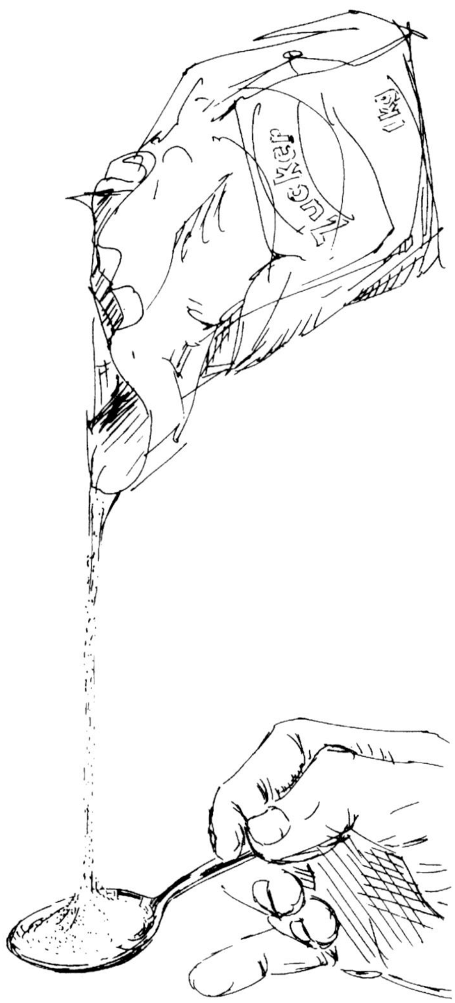

I experience an encounter with physical substance.

effects from them. On the other hand we can look upon the presence of chemically defined substances as a *condition* for the appearance of particular phenomena at the life or psychic level or see traces of these phenomena in the physical body, for the organism has usually formed them itself. The meeting of percept and thought experience takes place at the various levels of life processes (etheric body), soul (astral body) and I experience, in ways that are different from those of the physical. As regards substances, we are dealing with various realities (certainties) that occur in a particular way in each concrete case. Therefore, the *connection with corporeal experience* is different in each case too. So when Rudolf Steiner speaks about a 'flowing in of earth substances' into the human body, we should note that although a person's various higher members work in their physical body, their substances are of a different kind in each case. Let us make this clearer with the following exercises:

Sugar example

1. I take a spoonful of a white, granular substance, look at the form of the crystals and may already decide that it is sugar. Then I take a lick and find that my judgment is confirmed by the sweet taste. I also note its hardness with my tongue, experience its resistance with my teeth, and feel its grittiness. The sweetness increases in proportion to the amount in my mouth.
   Sugar is something quite definite for me at this physical level.

2. As the touch percepts disappear and the sweet taste spreads out in my mouth, percepts of a different kind appear: I feel a pleasantness and I am as if 'drawn in' by it. As the taste intensifies the feeling may reverse and become unpleasant or almost bitter. When the sweetness dies down my sympathy can increase

again and a desire for more may even arise. The whole process changes when I eat a second and even a third spoon of sugar. I become less and less inclined to taken another one and eventually turn away in disgust.

The fixed judgment 'sugar' can do nothing vis-à-vis this reality. Whereas if we direct our attention to the way the percepts change we feel that the instinctive urges of our soul body are continually changing. There we experience a rhythmic change of direction, which we accompany with the soul although it takes place relatively independently of it. This activity is a perception of the etheric body's living substance, the substance of life processes. We are usually only dreamily aware of this, if at all. The astral and etheric bodies are much more closely interwoven with cognition here than when we are faced with perceiving and having ideas about a physical substance.

In this way I experience the living substance in a life process.

3. Meanwhile I notice something else. I feel the saliva flow that continues for a while after the taste dies down or suddenly dries up. More inwardly I experience a strengthening after previous exhaustion, an awakening, or a pleasant sleepiness like that of a child when given sugar water to help them go to sleep. What I previously called sugar is now connected with my life body in such a way that I perceive my feeling of life, freer or more bound. So I no longer perceive the sugar or the life process in connection with my likes and dislikes of sweetness, but notice an ability to unite my feeling with the body (in will) or with the spirit (in thinking). Such feelings are more inward and complete than the previous ones. They separate themselves from a direct connection with intake and mediate between life processes. If I direct my attention at them I perceive the extent to which the conditions for living in feelings are present in my body.

Thus I experience the substance of my soul.

4. Finally, I completely lose the outer connection with what was ingested. It has become united with me. I observe that I can work freely with my limbs and that I experience myself in this activity that seems physical. This results in warmth. Here the warmth that lives in my own impulse for an action becomes sense-perceptible warmth, and the action becomes an expression of my personality.
   It is general knowledge that the presence of sugar in the blood in the form of glucose, which the organism makes itself and keeps at a particular level, is necessary for a free action of the I, and that in muscular activity glucose is degraded and carbon dioxide is expelled in the breath. In this last statement, I face the separate parts of my organism and its functions in the same way as at the outset with the sugar. Blood sugar becomes the outer world for this viewpoint. This obscures my view of a very differently constituted substance, one permeated by reality, which must be experienced in an entirely different way. It constitutes a whole that cannot be divided into parts. This whole is the I-bearing substance.

I get an inkling of the substance of my own being in the unfolding of will in my body.

So in eating sugar we find four levels of perception to which we have different relationships and to which we have *direct perceptual access*. We adjust our way of cognizing accordingly. Substances change and manifest differently at each level. We can summarize this as follows:

1. – Sight and touch perceive a physical object.
   – When we taste the sweetness of the substance the object disappears.
   – The fixed concept 'sugar' gives me certainty and a delimitation on which I support my self-consciousness.
   – In this relation to the world, I experience reality as an I encounter with *physical matter*. This is the conscious side of what happens in physical digestion: my *I* confronts the strange

foods in order to overcome them.

2. – The sensation of taste changes in accordance with the whole situation in which something is perceived: initially, later, on repetition, etc. The nature of this change now becomes the content of perception.
   – Connections arise between the soul and the substance of life processes, which change and which I perceive as a relatively independent activity. The initial desire and attraction turns into dislike and even revulsion and can rekindle later.
   – The astral body's rhythmically flowing instinctive perceptions are its reactions to living substance.

3. – Sense perception and direct desire disappears.
   – I perceive that I feel well or unwell and a feeling of exhaustion (deficiency) or refreshment (balance) stands in a more distant relation to direct intake of sugar.
   – But these percepts are only the background on which I experience the reality of ensouled substance: the ability to perceive and experience the vividness of the world.

4. – Self-consciousness (I-bearing substance) in the identification with my physical body, whose limbs I can use freely and can also observe from outside.
   – The physical body becomes an expression of the personality, which forms its own substance and then destroys it again through activity.

Cooking salt example

Observations of *cooking salt* can be differentiated in the same way. But the approach is characteristically different; not so pleasant, more 'formative'.

1. – *Physical substance*: I am dealing with salt. It might look like sugar at first glance. But the shapes of the crystals are quite different. I notice the cubic form and the fact that it cleaves at right angles. Just by licking salt I sense that the saltiness is very different from sugar.

2. – *Enlivened substance*: In some situations I may have a need for something salty. But pure salt is soon too much for me. I seldom desire it as such, but mostly only when it makes foods tasty and does not become obtrusive. Saltless foods taste bland and often monotonous. Thus at the level of living substance, salt has mainly a mediating role in the ingestion of food.

3. – *Ensouled substance*: when food tastes good, I, with my soul, am entirely engrossed in its special quality. If it remains insipid, I am not really interested in it.

4. – *I-bearing substance*: I feel that in my physical body I am free and that I can handle the situation. With the ability to place my self in three-dimensional space, I have made the right-angled, ordering power of the cubic crystal of salt entirely my own.

Via soul experiences (astral body) we become aware of *life processes* that convey the sense world relatively independently of the soul and yet in close connection with it. This occurs through our relation to substances in these observations of sugar and salt; from the experience of an object (physical

substance) via observing in the soul what follows the sense percept.
At this level of attentiveness, the percept cannot be firmly established because the life condition changes as more and more substance is absorbed. In self-observation we become aware of how our life unfolds as we swim in the waves of life processes.
The step to *ensouled substance* takes us further inside. The experiencing process turns round. The soul now finds itself in the changing pictures and life processes of the outer world, which are given above all by plants.
Finally, at the fourth stage our gaze is directed at an *identification with physical substances*, i.e. at how our own being participates in the inner nature of mineral substances.

Comparison of the Transformative stage of Sugar and Salt in view of their different effects on the Human Organism

The exercises allow us to trace how earthly substances are gradually taken into our bodily organization. Our relationship to them is traced stepwise from our experience of the object via sentient etheric and astral subordinates to self-experience in our own activity. Here, in 'tasting', we perceive them in a characteristic way as different (sugar does not taste the same as salt).
The inner connection between the qualities which we experience at the four levels of the physical, the living, the ensouled and the I-bearing substance is illumined in the phenomenon by the living picture that we receive of its origin and its position in nature as a whole and which leads to a kind of an inner gesture.
Through the stepwise transformation of these substances in the human organism capacities are liberated that become comprehensible from the way they are formed and manifest in the context of nature, i.e. from the 'phenomenal halo' (Manfred von Macken-

sen) which characterizes them as physical *substances* in the context of minerals or plants.
Mineral substances like cooking salt are very different from plant substances (e.g. yarrow flowers), because salt stands in the context of the whole earth's minerals whereas a plant has its own context, each species in a somewhat different way. The more plant substances are kept in their original context the closer the connection between the picture of the plant and the soul capacities that are liberated in the human organism. Substances taken from live plants and still showing an imprint of the life of the respective plant are more closely connected with the building-up of a person's soul life (effect on the astral body), whereas mineral ones are more connected with their I organization.
Most of *salt* as a mineral substance is dissolved in seawater. In salt deposits it is in cubic crystals and is extracted by dissolution in water and recrystallization.
Salt cannot be used as food but, when sprinkled thinly, it mediates the potential of nutritional qualities to taste stronger. The outer process of dissolving and combining corresponds to the human faculty of forming ideas in the soul that refer to earthly things, i.e. forms in the mind that accord with space, and of dissolving them again. This ultimately gives the I the ability to position itself with its body in physical space.
Sugar is formed in plants with the aid of light from carbonic acid from the air and from water in the ground. As dissolved substance, it is free to move around in the plant, or it can be deposited in starch and used later to build up and support the plant's earthly form (in the form of cellulose). When individual substances are isolated or 'purified' and in the same way as salt crystallized – indeed crystallization is a direct method of separation – they become closer to mineral nature. To a large extent this is possible with sugar which indicates that sugar's effect on the human organism is connected with the I. So sugar is quite similar to mineral salt.
Through the intake of this mineral-like plant

substance into the human I organization the will arises to direct the process that is the reverse of the one in plants (degradative) in such a way that the physical body can be taken hold of by the I and can become the expression of the personality. Salt and sugar bestow on the human I the mineral capacity of spiritually ordering, forming. The cubic crystallization of salt in the context of nature shows what the human being grasps from within in experiencing the dimensions of earthly space. Sugar is produced in plants. It gives the I the ability to become active out of the spirit just as the plant takes on shapes in manifold ways through the interplay of cosmic and earthly forces.

The conditions of production and the gestures that lead over to a human organism cannot be seen from the results of mere analysis. But they constitute an important connection with the body's requirements. To some extent, we can determine whether a particular amount of substance that is obtained from or given to an organism is a requirement for the correlative appearance of a capacity of that organism, but we cannot determine the inner connection.

The completely pure 'substance' or the isolated active principle is a postulate that corresponds to a particular way of thinking. And the concept of 'impurities' that cannot be removed is also a postulate. For even if nothing of the kind is demonstrable on the path of analysis, the fact of the existence of the path, i.e. the way the plant grew and the procedural steps in separating the substance from the whole or in synthesizing it 'from outside' still remain.

The closer we go to 'purity' the more the connection with the particular plant species is lost. Then the connection with its origin in nature becomes ever more general, but it remains an organic one. This is no longer the case with synthesized substances. In these, only the 'model' that is given in the 'structure' of the substance remains. The question that arises for me here is: to what extent can I compare the effect of such substances on the human organism with the effect on our perception of synthetic flowers, or with the way in which animals react to dummies?

I have occupied myself at length with what happens in the effort to prepare homeopathic remedies. . . .
I always saw the opposite effects as being present in the medium, in the ointment or whatever medium was used to incorporate the homeopathic pulverized [comminuted] substance. This medium takes on a different configuration, just as I become a different person in the eyes of society when I make the transition from wealth to debt. Substance makes the transition to its opposite state and then contributes this opposite state, which was formerly within it, to its surroundings. In this sense, I might say that while a substance possesses certain qualities as I reduce it to ever smaller quantities, as I approach a certain null point it acquires a different quality, the ability to radiate its former qualities into its surroundings and to correspondingly stimulate the medium with which I am treating it.

Rudolf Steiner: *Geisteswissenschaft und Medizin* (GA 312), Dornach, 5th edition, 1976, lecture XI, 31st March, 1920, p. 212; trans: in *Introducing Anthroposophical Medicine* (Spring Valley: Anthroposophic Press, 1999) I, (*Foundations of Anthroposophical Medicine*) pp. 155–6

Concentrated versus Diluted Substance as Polar Experiences in Tasting. Aspects of Potentizing

In the previous exercise, the directly recognizable effect of particular substances like sugar or salt is tasted. A specific effect on various parts of the organism at various levels is traced.
If we dilute the same substances in water so much that we cannot identify them by taste anymore, we then no longer taste the substances but instead the various qualities of the water used. (It is characteristic of the particular substance that to do this we need much less salt than sugar, and even less citric acid.) The water samples are then described, for example, as:

1. sweet, soft, with character
2. sulfurous, leaving an aftertaste
3. cold, refreshing, dry
4. insipid, non-individual, bland

Here I would like to recommend that readers try to perceive such differences themselves before reading the following assignments and explanations.
Through these expressions arise pictures of the respective water quality. An effect that has become united with the water that we drink, and that more

or less stimulates our organism to refresh itself, becomes communicated to us pictorially.
These indirect effects are much finer and spread evenly through the whole organism, unlike the direct effects of concentrated substances.
The properties numbered above correspond to: 1. sugar, 2. salt, 3. citric acid, and 4. the water that was used for dilution, without additive. Substances 1–3 disappear, as it were, into the 'surroundings' of the water that works as a whole. Water sample 4 is not fundamentally different from the others. With it the 'surroundings' themselves, so to speak, are tasted, i.e. the quality of the spring from which it came.
Contrasting 'tasting of concentrated substances' with 'tasting water quality' gives an experiential approach to understanding how substances can be introduced into the etheric organization more 'pointwise', via the physical body, or more directly, 'peripherally'. Through this we can begin to understand the effects of potentization. Potentization of substances starts from the picture that through stepwise dilutions the effect of the substance is reversed. If we take the percepts described as an indication of this transition, it will lead to an observational foundation at least for tasteable substances. Rudolf Steiner refers to the possibility of observing fluorescent substances. In a study, Georg Maier and Klara Van Dorpe showed how colors change in a stepwise dilution of eosin. Moreover, the conclusion from such observations is that we cannot proceed in an abstract manner, but we have to find the appropriate potency for each substance. Thus we still have not found out how we could observe the transition to the astral and I levels in potentizing.

This stimulation can consist in directly evoking the opposite effect described here, but it is also possible that this countereffect is evoked only by making the substance fluorescent or phosphorescent during or after exposure to light. Then we have called forth the countereffect in the surrounding medium.
These are the circumstances that must be taken into account. Far from succumbing to mysticism, the issue at hand is to finally observe nature in action for once, to truly study its rhythmic processes as they apply to the qualities of substances. This is a leitmotiv for recognizing where the effects lie. If you potentize a substance, you will eventually reach a null point beyond which the opposite effects are present. But that is not all. On the route that lies beyond the null point, you can then encounter another null point that applies to these opposite effects. By passing this second point, you can achieve even higher effects that work in the direction of the initial ones but are very different in quality.

Rudolf Steiner: *Geisteswissenschaft und Medizin* (GA 312), Dornach, 5th edition, 1976, lecture XI, 31st March, 1920, p. 212; trans: in *Introducing Anthroposophical Medicine* (Spring Valley: Anthroposophic Press, 1999) I, (*Foundations of Anthroposophical Medicine*) pp. 156–7

VII Health and Disease
– Considered from the viewpoint of a Direct Meeting with a Person

Man with Human – Polar Encounters

The previous chapter can be summarized as follows: the *plant* meets us as a whole when we look at it in connection with its surroundings. We do not perceive the connections of its mode of formation with its meteorological and cosmic environment sensorially since we are not dealing with something that is spatial, but with something active in space. We bear as capacities in our soul the forces in the environment of the plant that work on it. That is why we can recognize their effects in the phenomenal picture of a plant, or we can form our soul by experiencing them, 'tasting' them. Such connections become manifest in the way that we approach a plant without thinking about it. We can confirm this by self-observation through many simple examples: by completing an incomplete form; by becoming aware of metamorphoses in a series of phenomena or, for example, by studying a branch from a tree. We see it not only as an object, but also realize how it grew on the tree.

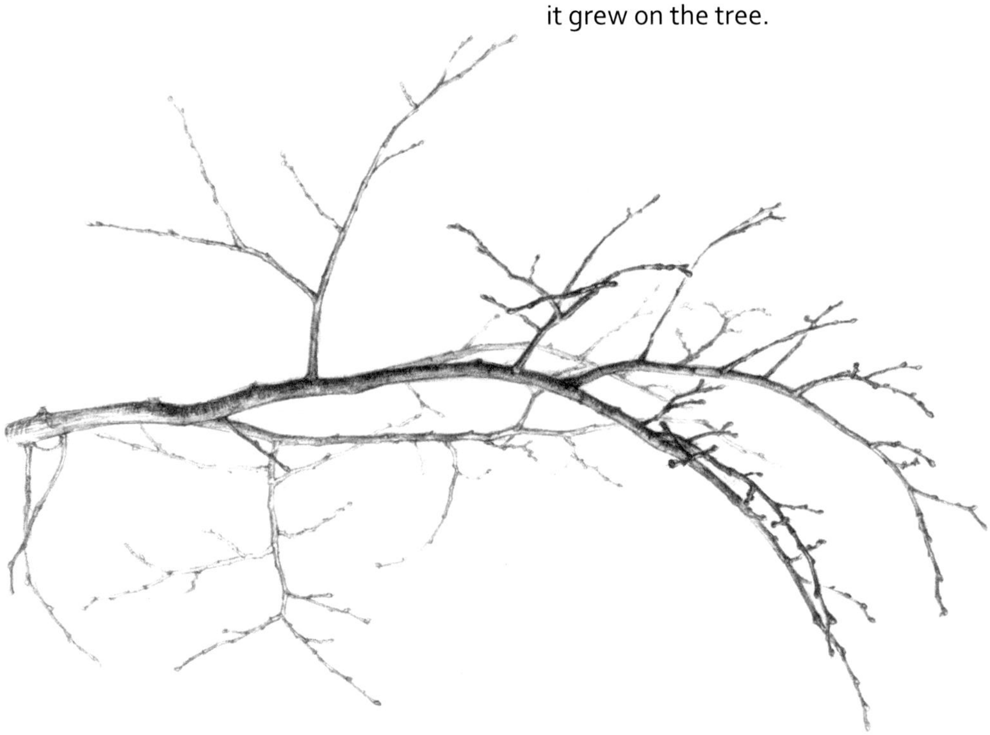

We can turn it to the correct orientation in space without knowing how we are able to. We obviously see this orientation not through a logical conclusion, but directly, as part of an intuitively grasped (unseen) tree connected with our upright body and from there with a pictorial *environment*, which we experience inwardly, and in which we are standing.

An encounter with a *person* is quite different from one with a plant. We perceive in them *inwardness*. We sense their ability to express themselves to varying degrees, their alertness or tiredness. In their blushing or paling we perceive a little of the innermost life of their souls and in experiencing it through them attune ourselves directly to it. These percepts refer to something independent *in the body* and in its surroundings. It is largely in accordance with these judgments of feelings that we manage our life with other people.

If we notice things that are not quite right in someone, we might ask them how they feel. When a doctor begins an examination that is usually directed at physical organs and chemical, physiological processes, his thinking becomes dissociated from an experience of the direct percept of the human being, and it turns towards the outer considerations of the body extended in space.

So that no misunderstandings arise we should add that the great value of scientific-diagnostic findings for medical practice lies in the 'control sample', i.e., on the one hand, in a search for how what was perceived in the first way becomes manifest in the physical body investigated from outside, and, on the other hand, in becoming aware of differentiated percepts of the way in which a particular individuality lives with their body.

Conventional scientific investigation discovers only outer things (physical organs) and the external processes of nature. This way we cannot tell why a particular process is like it is and not otherwise. We cannot even penetrate life processes with it. That is why in conventional medicine the basic principles of the human organism and its physiological processes are accepted as something that is given. Pathological processes are considered deviations from a norm. But such norms related to the physical body are *determined* through discoveries that, as 'experiences' in clinical investigations, were originally connected with percepts of a soul-spiritual kind and which become more or less unconscious assumptions.

Perception of Personality as a Starting Point for an Understanding of Health and Disease

In the following we will direct our attention towards the finer percepts that we make implicitly when we meet someone and which we understand directly as the modes of expression of his or her personality.

Let us examine this statement a little more closely. What do we perceive of a personality? In an encounter with it, the physical processes such as the mechanics of the limbs, temperature, rhythms of the heart and breathing, blood values, etc., do not stand in the foreground. As we said earlier, we experience it directly in the way it presents itself to us. For instance, in gait, gestures, tone of voice, speech, etc., we perceive a particular character and thus directly how this personality can become more or less fully manifest. In their unconscious resonance in us we regard the expressions of this person, for example, as restricted or as uncontrolled without being able to say exactly why. We feel that such a disharmony is a disease or a tendency in that direction. These observations show us that what we see of a person's soul-spirit aspect is present physically at a definite place and not, as with a plant, in the environment. We make mental contact with a person and face them in space when we talk to them. This relation induces us to look for an approach to understanding health and disease in the intuitive encounter with a personality, and after reflecting on the experiences, to proceed to the organs that *soul-spiritually* permeate the whole organism in various ways. A person's internal organs correspond to what we look for in the *environment* of a plant. With his organs, the human being is in a relationship with his physical environment via the life processes. The latter create a balance at various levels between inside and outside.

See threefold organization, p. 87.

The percepts that are involved here can initially be further differentiated in self-observation.

We obviously have something in us that allows us to experience the world in which we live as an ordered unity in which we feel included. In the preceding chapters we have already described how we can understand the formative forces that make connections in the processes of the human organism and other living organisms, and how to deal with them in thinking and imaginative activity. It is hard to describe the world of etheric forces. They are only perceptible in their effects when something concrete happens in the natural world around us. Light itself is invisible, but in perceptive thinking we become aware of the lighting context, as light ether, in which a particular object stands; or we notice the way in which the mode of formation of a particular plant follows the activity of the light ether.

We remain unaware of the action of our etheric body until something foreign is added. We recognize the activity of this principle in us only when we experience that there is something wrong with our organism.

The soul's dreamy experience of being in harmony with the world is already disturbed when we make a percept. Something foreign enters. We could describe this by saying that an awareness of something outside us arises. At the same time something becomes active within which counteracts what is perceived and that tries to bring it into harmony with us. An after-image arises in the bodily organization, and an unconscious sentient judgment arises in our soul. For instance, when we look at a color, something similar to the sensory experience arises that is experienced as the complementary color. Or when we look at a waterfall, a counter movement upwards arises, which appears as if physical-sensorial on the rock alongside. Mobile experiences of this kind in life reveal indirectly the harmonizing and healing effects and principles of the etheric body through the overcoming of the disturbance.

What works upon a sense organ from outside disturbs the etheric body's life like a stone that is thrown into a lake. The waves that arise stimulate the soul into quite particular experiences.

So becoming aware of something that is foreign to one's own nature is akin to a disturbance or disease, whereas the harmonization through a life process is akin to healing.

Disorders of the Senses

Even in a healthy perceptual process, each of the human senses creates a one-sided connection with the world that must be balanced again. When we see, hear, and orient ourselves we experience ourselves as a 'whole human being' in each case. It is only in this way that what is seen and heard can be experienced as a gesture.

Besides these 'normal disorders' there are also organic disorders when the soul, usually unnoticed, cannot put itself into a 'healthy' relation to the world.

We become intuitively aware of sensory disorders when we identify ourselves with another person in meeting them. In understanding the disorder, a picture of the healthy process arises indirectly in self-experience, which we apply to the disorder. For instance, we may notice from what a person says that they have a one-sidedness in their soul, an overly narrow or broad relationship to their visible surroundings. This relationship may show organically as short or long-sightedness.

Disease lies in the respective disharmony that exists in the relation of the soul to its environment via the body and that then has an effect on physiology and organ formation. The way a person's mind works determines their organ formation and vice versa.

Each sense comprises a particular relationship of the human being to the physical surroundings, which we experience as the outside world. Disorders in the senses can, of course, initially make our judgments wrong or inadequate, so that they do not agree with those of other people. But an intelligent consciousness can emancipate itself to a certain extent from this through a comparison with other percepts, and if it becomes aware of how it perceived.

Life Processes Connect a Person's Soul and Spirit with the World around

The life processes are involved in sense perception.

In an anthroposophical light, all that may be called a human sense which induces the human being to recognize the existence of an object, being or process in such a way as to justify placing its existence into the physical world.

Rudolf Steiner, *Anthroposophie. Ein Fragment aus dem Jahre*, 1910 (GA 45), Dornach, 3rd edition 1980, p. 31; trans: *Anthroposophy (A Fragment)*, (Spring Valley: Anthroposophic Press, 1996) Ch. 2, 'The Human Being as a Sensory Organ' p. 88

Note here how the inner nature of a physical organ is looked at from self-experience. On page 96 it says: We must keep the difference between the domains of the various senses clearly in mind, because we enter into a different relationship to external objects through each sense.

Correspondingly, we also have to look for an inner access to life processes and internal organs. A transition to this is provided by the sense organs that perceive things in our own body.

It is through them that the soul-spiritual part of a person is connected in a variety of ways with their body and the outer world. Life processes are in principle always rhythmic and their rhythms measured on an earthly scale can be very varied. We can observe how we participate in a life process without ourselves producing it when after-images form in connection with vision. For instance, when we concentrate on a red color something external is imprinted in the senses. Through a counterbalancing organismic change this is reversed in the inner life. From it, like a projection from within, we see a polar phenomenon emerging when we look at a sheet of white paper. It looks green for a while. In this whole process, consciousness can be modified from without in the picture formation and from within in experiencing.

The life process always becomes perceptible when it reaches its boundaries. Human life is supported by the respective rhythm and not by a rigid center. Its course is regulated by life instincts that give the soul impulses and that each time initiate the appropriate change of direction, for instance, to inhale after exhaling, and vice versa.

According to Rudolf Steiner (GA 45, Chapter 5) the course of life processes looked at from the point of view of the soul nature can be understood as follows: 'image sensations', particular needs, for example for air, food, etc., arise via the *life instinct* that determines the physical course of the process. This corresponds to what is known in ethology (animal behavioural science) as an animal's drive that at the moment of an actual encounter suddenly changes into 'desire' or yearning and ultimately into 'impulses for movement'. The life instinct then turns the life process in the opposite direction. Self-experience always lies in the middle of this swing of the pendulum.

See Jochen Bockemühl, 'Die Bewegung des Tieres als Ausdruck seiner Innerlichkeit', in *Elemente der Naturwissenschaft* **27** (2) 1977, 32–45

The process becomes pathological when the moments of consciousness in perception and experience (or in exhalation and inhalation), that happen at the boundaries, lose their inner connection with

each other.
Rudolf Steiner describes seven such life processes that come together variously in the human organism, namely breathing, warming, nourishing, secreting, maintaining, growing and generating.
The *'tasting' by various human organs* in the ingestion of substances can be viewed as comprising an interaction of life processes or a predominance of one of them. All life processes participate in each case. The respective 'composition' is the taste.
The perceptual capacities of inner organs (their tasting) that are mediated by life processes are not subject to conscious judgment by the I but work directly and are reflected as changing moods in the soul. Thus, relative to percepts of things, it is harder to stand back, so to speak, cognitively. They are experienced only imaginatively in their actual interrelationships.
In the following, I would like to go into life processes and then venture a cautious approach to the perceptual and experiential qualities of the internal organs. This study proceeds from our personality with which we identify ourselves. It lives and experiences itself in a center from which it faces the physical-sensorial world, on the one hand taking in and on the other hand presenting itself outwardly.

Life Processes that are Directed towards Intake: Breathing, Warming, Nourishing

Breathing
In *breathing* we can easily observe in our body how the urge to inhale certainly stimulates this activity, how we cannot stay with it for long and how it triggers a desire for the corresponding counter movement.

If the process does not happen properly, i.e. if the reversal does not occur in the right way (for example in holding our breath) anxiety and dimming of consciousness arise.
Rudolf Steiner categorizes as breathing the rhyth-

Life Processes:
Another type of activity plays into our sense life. Here, too, we can distinguish a number of different domains. The first to present itself is the process by which our body's inner life is supported from outside, namely our breathing. *In the breathing process, our bodily life touches the outer world. It sets itself against the outer world in a way that it cannot maintain, in order to receive from it the strength it needs to continue. These words more or less express that which can be said about what manifests itself to us about the breathing process without going into the results of sense perceptible science. The latter belong to the field of anthropology. What has just been described is something we experience directly in our own life – in our own need for air, in observing how life is inhibited by lack of air, and so on.*

Rudolf Steiner, *Anthroposophie. Ein Fragment aus dem Jahre*, 1910 (GA 45), Dornach, 3rd edition 1980, Ch. IV, 'Die Lebensvorgänge', p. 50; trans. (edited): *Anthroposophy (A Fragment)*, (Spring Valley: Anthroposophic Press, 1996) Ch. 4, 'The Life Processes' p. 107

mic process that take place between perception, experience, and ideation. Because it is followed more consciously it is easier to observe what occurs in it inwardly. Here are a few examples:

The seeing of *forms* is based on a desire for shaping. We grasp them with bodily senses (senses of balance, movement, and life) and identify ourselves with the form received in such a way that we can bring it out of our bodily movements by drawing it on paper, or allow it to become fixed in a mental image. We cannot remain in the finished form. We have to be able to forget the mental image again. Through a kind of digestion of cosmic influences of this kind via the perceived form the specific influences are transformed into a general, differentiated capacity of the soul to experience and shape things. From this the organism creates not only its soul form and but also the physical foundation of its shape in bones etc., so that it can unfold itself in it.

Seeing *colors* is based on a corresponding desire for an experience that brings a life process into a focussed etheric movement that begins to sound in the soul. We cannot remain in this either. The color impression dies down. But the capacity to experience, which makes the etheric body mobile in its interaction with the astral body, is enriched from within through the digestion of impressions.

Smelling is bound to very finely divided, aeriform substantiality. We notice a desire to place ourselves soul-spiritually into certain situations, from which deliberate will impulses can then arise. The digestion of these impressions leads to a further development of the astral body (life of instincts).

We have summarized seeing of forms and colors as well as smelling in connection with cosmic nutrition that shapes the human organs from within. Each organ is a whole from this viewpoint, so that we can speak of 'lung man', 'heart man', 'liver man', etc.

In connection with the life process of breathing, an understanding of these processes (the desire for form) also sheds light on a particular aspect of the human organism, namely the connection of *lung man with the constitution of the landscape*, i.e. the connection of the soul with a particular homeland in which the soul formed itself. An overly strong connection of our constitution with a particular landscape (e.g. homesickness) or an illusory relation to earth conditions (e.g. not being able to settle down anywhere) tends to be pathological.

See also Cornelis Bockemühl,'Der Bezug des Menschen zum geologischen Untergrund' Merkurstab, 5/1996, 382–391; trans: 'How Human Beings Relate to the Geological Base', *Journal of Anthroposophical Medicine*, 4/1996, 38-49

Warming

The life process of *warming* comes to expression in maintaining a particular body temperature that is independent of the environment.

Warmth formation in an organism is dependent on breathing and nourishing. It becomes physically liberated through conscious and unconscious activities.

We can warm ourselves in soul, be warmed from without, feel overcome (too hot), or cope with it (too cold). We experience warmth as something that is in between the bodily and the soul aspects. We are somewhat more distanced from it than our breathing is from the environment. On the one hand, we can warm ourselves up or cool ourselves down etc., and, on the other hand, we have in warmth, as an experience of our own power of initiative, a constantly renewing power of self-experience that also works from within outwards. Everyone is more or less able to live in this warmth element in harmony with the world and to have a sense for the 'open secret' such that he feels united with the rhythm of life. Thought experiences and limb experiences are in natural harmony. Whereas the sensory disorders described above become manifest in relation to the percepts of particular senses, here we are dealing with an *overall experience* with respect to the worlds of the senses and of activities.

This experience of 'being in the world' is also connected with every true experience of certainty.

If we try to grasp this kind of relationship between the human being and the world imaginatively, we

Regarding experience of the inner nature of physical organs:
Let us assume at this point that for inspired consciousness I were to see this material world permeated by the spirit world. Now what is perceived spiritually outside does not arise within a person pictorially, but rather we now learn to recognise the spiritual outside in the way it is reflected within the person. It is reflected as their physical organs, as lungs, liver, heart, kidneys etc., and as everything else that is at first material within them. It is a complete inversion or reciprocity. Whereas the material world is reflected in us in a soul-spiritual way for ordinary consciousness, the spiritual world is reflected in us through our organs. We learn to know ourselves inwardly as physical people when we become aware of the spiritual world around us. Until then nothing will be understood of the physical human being.

Rudolf Steiner, *Damit der Mensch ganz Mensch werde: Die Bedeutung der Anthroposophie im Geistesleben der Gegenwart*, 'Die anthroposophische Forschungsmethode', 10th April 1922 (GA 82)

reach a perception of a second organ of the human organization, namely of *heart man*, who is to be looked upon as being closely connected with the life process of *warming*. This organ mediates between breathing as cosmic nutrition ('above') and physical nutrition ('below'), and between 'warmth' from within and from outside.
Disorders in this heart man are manifested in the soul as anxieties and the like, because the aforementioned experience of the harmony with the environment does not come about at all or only occasionally.

Nourishing
With the life-process of *nourishing*, bodily life relates to the outer world via a desire to replace substances that have been used up. Longings for the intake of solid or liquid food express themselves in hunger and thirst. Another soul mood announces itself as an expression of an organ's perceptual capacity in the appearance of the need to take in specific substances, dissolve them and transform them into our own substances. This mood comprises our changing likes and dislikes for foodstuffs as well as our corresponding unconscious processes in digestion. So this perception regulates the life connection between the human organism and physical substances and their metamorphosis in digestion. Whereas hunger is connected with a 'desire for solid, for forming' (lung man), imaginative perception sees a connection between 'thirsting and tasting' and *liver man*. On the one hand cosmic nutrition participates from within in the development of this organ and its various requirements (build-up of the organ), and, on the other hand, the participation of the astral body is stimulated in various ways from without via taste.
When we make differentiations in this way, single observations and the results of laboratory investigations have a place in the whole or they provide a reason to correct the picture that we had formed through a perception of the personality. So organs that are experienced from within like this can be

characterized as diseased in that the soul's relationship with the world is either too open or too restricted (e.g. excessive thirst, bulimia or anorexia).

Secreting as the Central Life Process

The life process of *secreting* lies between the three life processes that above all serve intake (breathing, warming and nourishing) and the three that follow (maintaining, growing and generating) through which the human being places him or her self in the environment.
Here we are dealing with a bodily interaction between outer and inner: substances are secreted from the life of the physical body. They are either excreted entirely or they combine (e.g. as saliva, stomach juices, etc.,) with nutritive substances coming in from outside, bring them into line with the organism, render them suitable for absorption, and secrete them into the interior. This usually happens in different ways within the limits of the body, i.e. in the mouth, stomach, and intestines. Nevertheless we are dealing with a relationship in which the physical surroundings have their closest interaction with the physical aspect of the human body. Conscious tasting is especially closely connected with this life process. The route through the digestive organs from mouth to stomach and intestines gives an inkling of how this tasting usually changes and becomes ever more unconscious. Health expresses itself in digestive comfort. More conscious tasting in digestion or pain are signs of irregularities. Excretion of substances occurs at the end of the digestive tract, paralleling the interior secretions. Here too there is already a hint of the aspect of the life processes that are mainly turned outwards (maintaining, growing, generating). Illness manifests here in constipation and diarrhoea.
But secretion processes are also involved at other levels, as in sweat formation and internal secretion. Here we are dealing mainly with the organism's involvement with the water element, which is ex-

Until then we will get to know the heart, lungs and liver outwardly through anatomy, but not their connection with the outer world. Through anatomy and physiology we get to know the heart, lungs and liver as if we were getting to know that a person has all kinds of ideas within them without realizing that these inner pictures refer to the outer world. We do not understand that these organs are related to the outer spiritual world. This is the origin of what makes it possible for spiritual science to work in a rational medicine, for example. For only now do we really get to know the human being, become acquainted with the inner nature of the human organism. Previously we could not get to know it at all, just recognize it outwardly.

Rudolf Steiner, *Damit der Mensch ganz Mensch werde: Die Bedeutung der Anthroposophie im Geistesleben der Gegenwart*, 'Die anthroposophische Forschungsmethode', 10th April 1922 (GA 82)

Why Do People Fall Ill? . . .
Natural processes are known primarily from observation of the world outside the human being, and from observation of the human being only in so far as this is done in exactly the same way as observation of nature, outside. The human being is seen as a part of nature, one in whom the processes that can also be observed outside are highly complex, yet nevertheless of the same kind as those in outside nature.
But this is where a question arises that cannot be answered from this point of view. How do natural processes develop in the human being . . . that are the opposite of healthy processes?

Rudolf Steiner and Ita Wegman, *Grundlegendes für eine Erweiterung der Heilkunst nach geisteswissenschaftlichen Erkenntnissen* (GA 27), Dornach, 5th edition 1977, Ch. II, 'Warum erkrankt der Mensch?'; trans. (edited): *Fundamentals of Therapy* (London: Rudolf Steiner Press, 1996) Ch. 2, 'Why Do People Fall Ill?' p. 13

pressed in the kidney-bladder process.
If after the intake of solid or semi-liquid food, perception of and dealing with substance enter the foreground of experience, the connection between secretory and resorption processes of fluids with soul experiences becomes much stronger. We can look upon kidney-bladder processes as 'kidney man'. The kidney is a main secretory organ that lives in the alternation between concentrating and diluting. This 'breathing process' in the fluid organism is comparable with lung processes:

- *Lungs*: desire for solids, for forming;
- *Renal-bladder process*: desire for freeing the soul nature from a tendency towards solidification in the dissolution.

In kidney man, spasms, fear and apathy contrast as one-sidednesses with the hunger for air and movement and agitation. As part of this, in the broadest sense, belongs all the emotionality that is connected with hormonal processes.

Productive Life-Processes: Maintaining, Growing, Generating

Maintaining
Waking and sleeping also interact with each other through a life process, namely through the desire for activity and consciousness that alternates with the desire for rest and regeneration. This process can be looked at under the aspect of maintaining. It involves the conversion of physical into living substances (sleep) and from these back into the lifeless substances (waking). With this we experience being enlivened or tired. The formation of the immune system may be especially connected with the maintenance life-process too.

Growing
In the life-process of *growing*, the organism can extend itself, increasingly shape itself as a whole

out into its environment. In a healthy way, we feel ourselves strengthened as a whole in a body that expands. Here living substance arises out of soul substance, and through the latter a man works back upon his life processes. There is a close connection between *maintaining* and *growing*, for in sleep the physical organism can become a *picture* of the soul through the growth processes and can refresh itself. The consciousness forming of the I and astral body is not present.
During sleep in the healthy individual the formative principle (etheric body) that creates connections in the bodily organization is maintained in its orientation to the I and soul organization (astral body). If this connection with it is lost, rank growth occurs because growing then only serves the physical body. This kind of illness consists of an overly loose connection between the upper and lower members of the human being. Then the physical and etheric bodies become independent of the higher ones. Disordered growth or other life processes spread through the organism. During *waking life*, the healthy feeling of the soul is supported by the etheric organism. Consciousness arises through the formative forces freeing themselves in various ways from the etheric organization in thinking, feeling and willing and becoming soul experiences. Feelings of tiredness arise at the same time.

Generating
In the life process of *generating* the form is repeatedly brought to a conclusion anew. People feel themselves growing in their bodies in such a way that through it they assume a willed relation to their environment and transform it. Ensouled substance passes over into I-bearing substance and works back into the formation of bones, nerves and organs. When a living physical body repeats and separates in the generating of its own form, *reproduction* occurs. During waking life, spirit and soul suspend the normal activities of the body and subordinate them in such a way that they can work in the direction opposite

The idea seems to be that the spiritual, non-physical aspect of the human being has a complex natural process as its physical basis, which is like a continuation of the natural principles that pertain outside the human being. But simply consider – will continuation of a natural process based on a healthy human organism ever bring the mind and spirit alive? The opposite is the case. The life of mind and spirit is extinguished if the natural process continues in a straight line. This happens in sleep; it happens in a faint.

Rudolf Steiner and Ita Wegman, *Grundlegendes für eine Erweiterung der Heilkunst nach geisteswissenschaftlichen Erkenntnissen* (GA 27), Dornach, 5th edition 1977, Ch. II, 'Warum erkrankt der Mensch?'; trans: *Fundamentals of Therapy* (London: Rudolf Steiner Press, 1996) Ch. 2, 'Why Do People Fall Ill?' pp. 13–14

Thus even the normal interventions of astral body and I organization in the human body relate not to healthy vital processes but to pathological ones. When spirit and soul take effect, they cancel out the normal way the body is constituted; they change it into its opposite. With this, however, they put the organism on a road where illness wants to begin. In ordinary life this is regulated as soon as it arises by a self healing process.

Rudolf Steiner and Ita Wegman, *Grundlegendes für eine Erweiterung der Heilkunst nach geisteswissenschaftlichen Erkenntnissen* (GA 27), Dornach, 5th edition 1977, Ch. II, 'Warum erkrankt der Mensch?'; trans: *Fundamentals of Therapy* (London: Rudolf Steiner Press, 1996) Ch. 2, 'Why Do People Fall Ill?' p. 16

to that in sleep (degradation versus building up). In general this is the picture of the beginning of the diseased state, which is normally eliminated likewise by the etheric body (self healing). Disease comprises the prevention of this self healing.

In this sense the life process of *generating* works in various ways in the members of the body. The thinking process is the most separated from the physical organism. For instance, whereas sensation is directly connected with sense perception, a human thought is not in a living interaction with an observed thing or process. Because the I identifies itself with things by *unconscious* willing, as we showed earlier, in *consciousness* arises the situation of facing a thing as something separate. Therefore we can also see a death process in healthy, clear cognition.

However, there is a living interaction between the I organization and the limbs. The movement of a limb proceeds from an intentional idea (perhaps of what is to be taken hold of). This dives down into the unconscious but then immediately lets go again. Then the arm and hand make the corresponding movement that is accompanied by sensations of the senses of touch, balance, life and movement. Free, conscious experience is an expression of the astral body. If pain arises consciousness is intensified and feeling acquires a content that more or less clearly refers to a particular physical organ and binds it to it. If a movement is accompanied by pain it is much harder to perform and we are much more aware of it. Our mind has to work a lot harder. The soul connection to the physical organ now becomes a compulsive, direct connection with the physical body. Therefore paralysis can be regarded as an intensification of this process. From this viewpoint, the healthy movement of a limb is the beginning of a paralysis.

If the I dives in completely and cannot let go a real paralysis arises in the same way as, in a judgment, something definite is separated from its connection and held fast.

The limb becomes a thing in the outer world. What is right and proper for bones and organ forms has a deadening effect on movement. From this viewpoint disease signifies an overly close connection between the astral body or I organization and the physical organism. In which case healing must comprise the right separation of the soul and spirit from the physical organization.

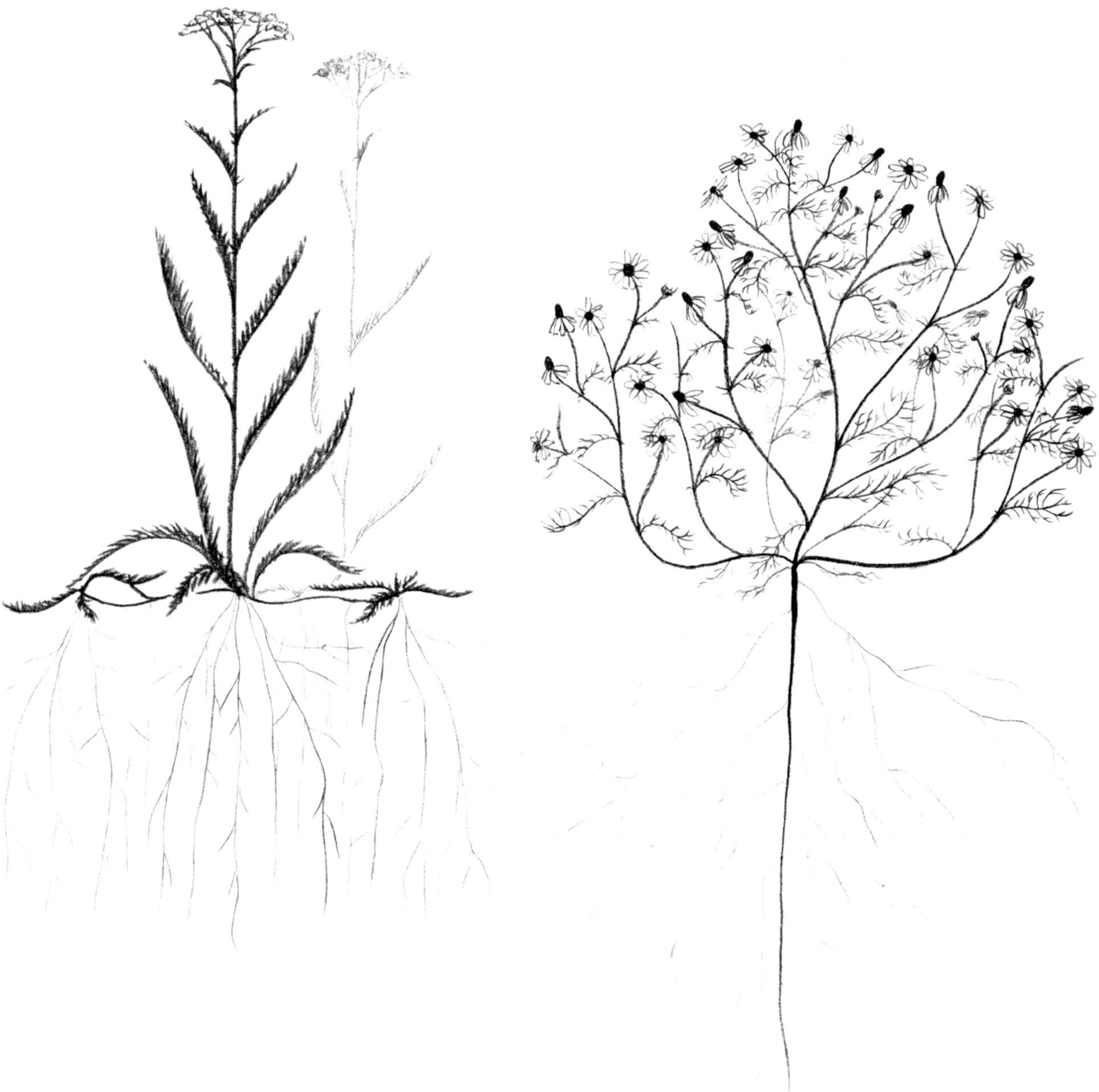

VIII Two Examples of Plant Observation: Yarrow - Chamomile

What follows is a description in greater detail of a more concrete approach using the examples of the two healing plants we have frequently referred to above.
Comparing them and letting them shed light on one another has proved to be a help in making their respective properties more visible. This, more biographical, approach is intended to show how important it is for everyone to find their own way into the connections sought. We can only point the way for each other.

A Comparison of Yarrow with Chamomile

A reasonably good sketch of *yarrow* with its erect, compact, delicately pinnate leaves immediately identifies the plant for someone who knows it, even if little detail is visible. Anyone using *chamomile* in healing or in agriculture, gathering and preparing it, also has a very clear picture of it. But, unlike yarrow, this green herb, with its open, threadlike leaves can, until its flowers open, easily be confused with similar species at field margins. Many people have a much more specific picture in connection with the smell of chamomile, its dried flowers, tea, or possibly even a steam bath. Such experiences form a kind of intuitive anticipation of the inner nature (being) of both plant species. Although we do not know exactly what this is giving us, the anticipation creates an inner orientation for getting to know more about it. The clarity and mobility of the perception is dependent upon how familiar the observer is with yarrow or chamomile. The perception is simply there with what is seen or smelt, although it does not become fully conscious or it immediately changes into a mental image of the object. Now it is a matter of accessing the realm that lies *before* the mental image, from which it first forms, in order to be able to build up the perception anew ourselves in the way we did with the preceding exercises.

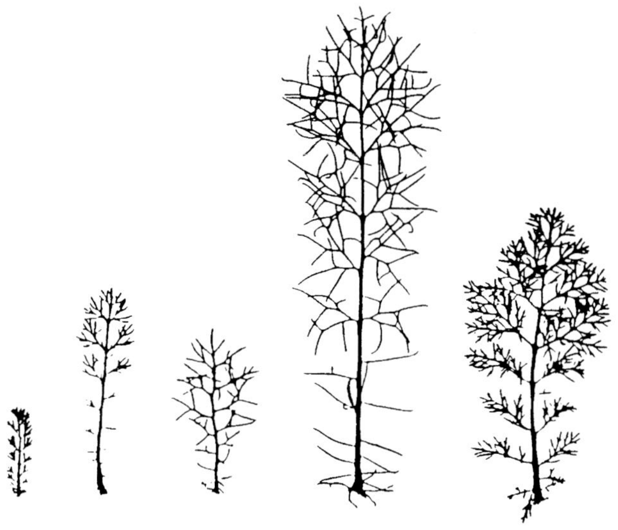

Rosette leaves of chamomile and similar plants, from left to right:
Rayless chamomile *Matricaria discoidea DC.*
Roman chamomile *Chamaemelum nobile L.*
German chamomile *Matricaria recutita L.*
False chamomile *Matricaria indora L.*
Stinking chamomile A*nthemis cotula L.*

Ingrid Pleier: 'Ein Weg zum Verständnis der Echten Kamille über den Vergleich mit anderen Kamille Arten', *Elemente der Naturwissenschaft* **60** (1) 1994, 9–31

VIII Two Examples of Plant Observation: Yarrow - Chamomile

A drawing exercise: plant looked at briefly and then drawn blind. The gesture is clear.

Drawing one day later from memory

Drawing from memory under various conditions helps us to do this. Drawing details at the scene trains exact observation. But it also fixes the mental image.

One way of training exact perceptivity that is kept mobile is to observe something exactly and then draw it when we are no longer on the spot, perhaps the next day. The mental image becomes more mobile if we try to reconstruct on paper, out of the mood of the whole, the impression of the details that were forgotten in sleep. It is formed from its relation to surrounding phenomena. In addition we can begin to become aware of what we spiritualize with these activities.

Another thing that helps is to look for a plant at various places in the landscape and at various times of the year in order to get to know its manifold formative possibilities and relationships. Here we try to observe the details and their connections, and we pay attention to what we actually perceive in each case and to the factual details that we gradually add to it.

Yarrow in Summer

In many places, yarrow begins to flower in June. In each place it takes on a different shape. On dry, close-cropped meadows, it does not grow higher than the width of a hand. There its leaves are narrow, pointed like an awl, and so dense that the fine pinnations that they comprise can only be distinguished close up. Its flowers are a little easier to distinguish (a).

In contrast, in a shadier place where fertile, cultivated fields are bordered by a ditch, yarrow grows to above knee height. The leaves spread out. Here the little pinnations look like a delicate, open network. Nevertheless, the contours of the overall leaf forms remain quite distinct (b).

A particularly dark green patch stands out in a relatively dry, open meadow situation. Here the grasses and herbs are taller than others of the same species nearby and their forms are all firmly erect, and not somewhat swollen and limp.

Amongst them there is also a yarrow plant with large and very delicate pinnate leaves. The ashes from a wood fire have been left behind there. We see something similar at places where a mushroom circle used to be (c).Even at places where yarrow grows luxuriantly, it can still develop its finely structured shape, give firmness to the substances it forms, and hold in its outlines so that it looks compact.

Richly flowering yarrow in a multi-colored stand in an open south-facing position beside a farm track next to a grain field; marshy ground. Omang Farm, Norway

It flowers until October. On picking its inflorescence we feel the rigidity, and as the stem is so tough, we have to take care not to pull out the whole plant. Our hands soon become resinous and sticky, and the fresh, aromatic-herb smell that yarrow emits only when it is rubbed becomes increasingly distinct.

In the meadowlands of Ticino near Milan, rice has been irrigated since Leonardo da Vinci's time. This leaves large tracts of water with vegetation exposed to the sun. These regions are oppressively hot and humid in the summer, and there is so much fog all year round that no one would at first expect yarrow to grow there. Nevertheless, we found a yarrow plant, in the middle of a track between rice fields and far from the edge of the woods, on a farm whose owner thought that yarrow did not grow there. Its being
there at all demonstrated the sort of conditions under which it is possible for it to grow.
So the natural occurrence of a plant can indicate specific qualities of a place. Conversely, the 'moods' of landscapes in which yarrow occurs widely can make the characteristics of this plant experienceable.

Yarrow at the edge of a track on a highland pasture with sandstone soil. Osterdalen, 'Fjellvang, Kappenkjelen', Norway

Yarrow at the edge of a track in a green stand; partially shaded, good soil under cultivation. Alpine farm in Stange, Norway

Chamomile in flower in summer

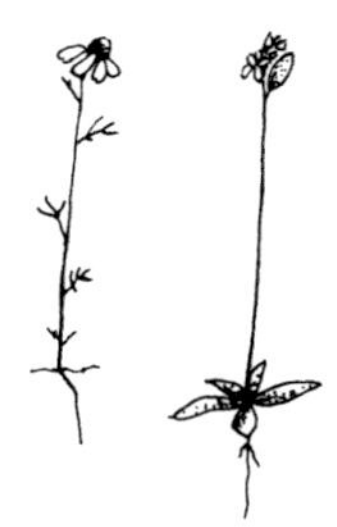

Chamomile compared with common whitlow grass, *Erophila verna L.*

Chamomile in summer

I have to search for other places to encounter chamomile: arable fields, often between growing grain; on the edges of farm tracks or on field margins; at mud encrusted stony places. There the smaller, loosely branched, multi-flowered bushes shine in the June sun. The green is lighter and more delicate than in yarrow. Lower down it is starting to wilt and pass over into beige. The multitude of flowers floats there like yellow-white stars. Flowering usually begins in May and dies down earlier than in yarrow.
In Hungary's treeless plains chamomile covers large surfaces of land where no grass can grow because water stands in puddles on encrusted ground after heavy rains. Close inspection reveals tiny chamomiles that one can hardly recognize as such.

Yarrow in winter

My observations are quite different in the wintertime. I find a desiccated, snapped yarrow stem with seed capsules. Blackish and shrivelled leaves hang down. I can still see how upright and straight the stem once grew. I must use my memory to help me to recognize the plant properly; its leaves can be reconverted into dark-green, erect 'feathers' in the imagination; I can also reconstruct the inflorescence. In this way the picture of yarrow in flower becomes ever more distinct. Although it cannot now be seen, it belongs to the full reality of what is now observed. The seeds fit tightly into their dry capsules. When ripe they are somewhat flat and extended. What do they have to do with my memory picture of leaf forming, sprouting and flowering yarrow? The fact that I harvested them is the only thing that at first connects them with it. A new, perceptible connection would only be created if I let the seeds germinate.
A tiny rosette is growing on the ground next to the dry stem, which shows that the yarrow did not completely spend its energy in seed formation but is sending out side-runners well below the surface

Yarrow in winter

Yarrow seeds

Yarrow remedy

of the ground.

Now I take a sniff of the dried flowers as if I had never smelt them before. The smell is very distinct, but hard to describe; hay-like, somewhat bitter, pungent and yet delicate. The effort it takes to describe these impressions helps me to consolidate inwardly the fleeting experience of the aroma. A 'reticent, insipid' experience arises.

I follow in my own body the taste of the remedy prepared from the inflorescence and upper leaves. First I feel in my mouth, as if from outside, something straw-like. A delicate, bitter taste emerges from this. As the bitterness, the tanginess in my mouth gradually dies down, my stomach becomes noticeable through a 'strengthened' or 'tightened' feeling. After this I can no longer distinguish between what is 'I' and what is my perception of something foreign. But my bodily condition has changed in such a way that I experience it to be in an imaginative connection with the way yarrow is formed and with the mood that is brought on by the aroma.

Chamomile in winter

In winter at the edges of fields we find very small rosettes with delicate, threadlike, gently curving, pinnate leaves which began to develop from seeds in the previous autumn.

Chamomile seeds are smaller than those of yarrow. Under a magnifying glass they look slightly bent like horns. Sometimes a dried up, straw-yellow flower is still clinging to the blunter end. If I bite into one of the tiny seeds a fatty texture is detectable, and the characteristic chamomile taste appears.

If I want to include the summer observations in order to get at the full reality of the live seed, I rarely have dried remnants of last year's plants to aid my memory, as with yarrow, because they have already rotted. From dried chamomile blossoms emanates a familiar, rich, somewhat honeyed, bitter aroma and I am well acquainted with its expectorant and anti-inflammatory effects in steam baths where the

'intrinsic warmth' of the aroma is intensified by the heat. The organism defends itself if the aroma becomes more intense and penetrating. It is a matter of enduring this sensation if I want to be healed.
The taste of chamomile tea can also be traced in the organism. I am looking for the sweetly-pleasant, settling effect that radiates down into my stomach and intestines when I need it for stomach ache – otherwise I sense it is somewhat insipid and unpleasant and thus refuse it.
Such exercises show me that it is the 'yarrow' *substance* and the 'chamomile' *substance* which I approach indirectly, through all the subtleties of the way they are formed, and directly in smell and taste. The indirect approach through the way they are formed, with which I worked before, gives 'outlines' to the direct approach based on my experience of smell and taste, and lets the still very indistinct percepts of effects on my bodily organization become more pronounced.
In all of this I already understand and experience a little more of the power that is given to us with these substances in nature. I can go one step further to clarify this.

Chamomile seeds

Chamomile remedy

Building up a symbol of greening yarrow

Stimulated by such experiences we can make use of drawing, and by entering into detail we can form initially separated perceptions of the different stages of the plant's development: greening, flowering and fruiting. Then we can compare these and also consider the way the plant above ground is formed in connection with root development.

For instance, it becomes increasingly interesting as we go into the details of a yarrow leaf with the aid of drawings. At first it leads to estrangement (if we use magnifying glasses and microscopes, for instance). Completely new phenomena arise against which the plant's original picture pales. We study the irregular, mobile branching of the pinnate leaves which nevertheless have a very regular, closed overall form. How can we find their context?

People might not recognize as yarrow an image of a very much enlarged part of a leaf. The outer connection would become clear if we were to trace the route by which the image was made, it is true. But it is not enough to reduce the leaf again in the imagination and to put it back at the right place on the plant so as to insert what was newly perceived into the picture of the plant. This is only *one* aspect. The perceptual impression changes with the change in size. I get a little closer to the way that yarrow lives vegetatively and develops out of this when I trace the particular shape of the fine branching of the pinnate leaves and follow up the way in which the overall form results from this. Then the 'reverse ordering' of this inner experience presents an extension of perception, clarifies. The picture obtained is looked for in the other details which involves paying attention to how vegetative processes occur in relation to the whole: in the branching of the pinnations of the leaf; in the inflorescence; in the spreading of fine roots and runners, etc. (These details do not need to agree with one another – as in the present case, they could also seem contradictory, until we have discovered their complementary character.)

We can sketch a yarrow leaf as if we are building up a symbol, that is, with a feeling for all the things in a greening plant that we can become aware of. For instance, I notice the way in which:

- the leaves of the rosette push up like an awl of densely-packed feather tips, increasingly unfolding in space one after another;
- the plant with its underground runners spreads out into a number of rosettes and thus forms a closed carpet;
- the flower shoots rise from this carpet in summer and become branched only near the flowers;
- the metamorphosis of leaves runs somewhat like that in plants with a closed leaf surface, i.e. with spreading and contraction, for they display a clear contour below the pinnate tips;
- finally, the characteristic taste in the upper leaves and inflorescence increases towards the top.

All of this can consolidate itself in our experience and can be summarized in a symbolic idea of a single leaf. All our experiences of greening yarrow now become concentrated in this.

Poppy, single flower

Yarrow inflorescence

We usually see the plant in flower as a whole, and we have a much more specific experience of this than in a greening plant.
For instance, we need only think of a poppy flower; there is no longer any green in it and the metamorphosis has been completed. The non-green color, i.e. red, leaps out at us; it contrasts with the dark, black-violet inside. As an overall phenomenon, this picture is something completely different from the one we get from the development and metamorphosis of a green poppy plant.

In yarrow the flower is very complex. If we want to find a single flower, we have to look at it quite differently as we go up the plant from the lower green part, via the inflorescence:

1. first we meet the vegetative branches that are coming to life anew in the inflorescence;
2. here there may still occasionally be a few small, finely tipped leaves concluding the metamorphosis of the green leaves;
3. next we come to the green flower-heads in which the plant's somewhat bitter taste is concentrated;
4. then follows the real flower level;
5. we find little five-rayed flowers; these are really whole flower heads, white, or perhaps somewhat reddish with yellowish pollen;
6. only then do we arrive at what a botanist calls a 'flower': mostly five marginal ligulate florets raying radially round a group of tubiform florets which in this phenomenal picture take on the position and function of the inner organs of a single flower.

Yarrow inflorescence

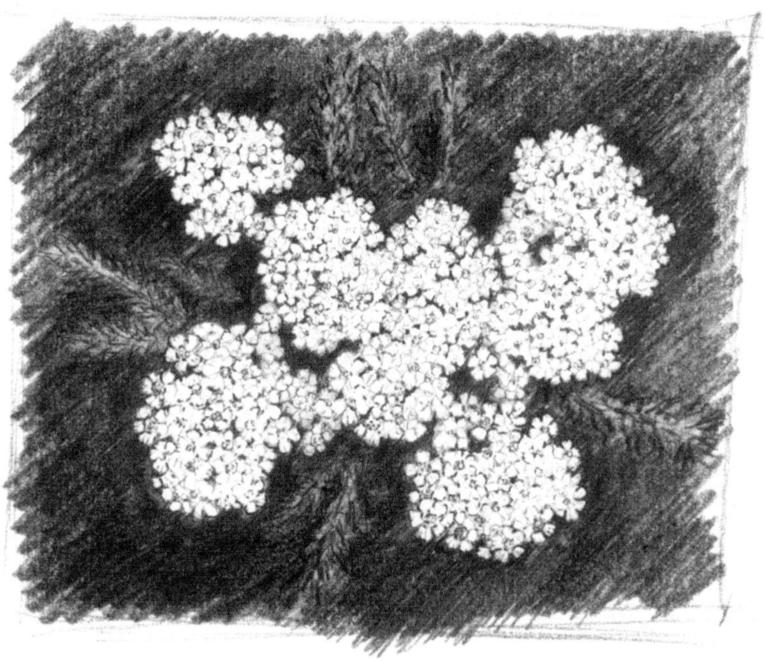

Above: flower heads
Below: single flower of yarrow

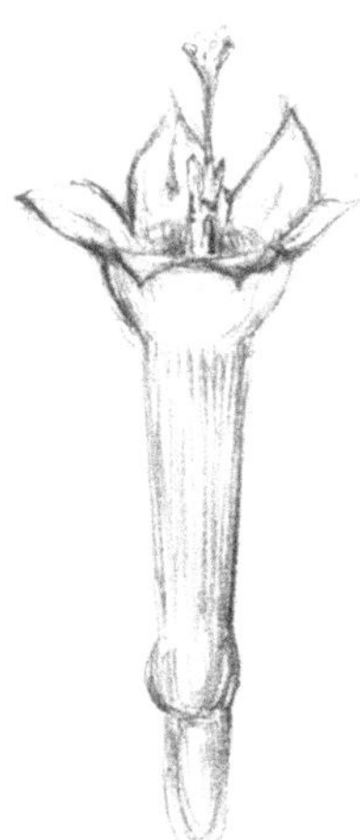

These composite flowers do not appear singly at the end of the main shoot as in sunflowers. Nor do they radiate from a center as for example, a *single* flower in a hogweed (*Heracleum sphondylium*) umbel, in which the bilateral symmetrical development of the florets is orientated towards a center. In yarrow, by contrast, what we saw in the leafy part becomes concentrated higher up: the fine, mobile branching and the intensification of bitter-aromatic nature goes right up to the flower level where the composite flowers are grouped as an overall inflorescence. The many little flower heads of yarrow seem to stop in a single plane the stream that comes from the vegetative region below.

Thus what gives the leaf an invisible outline holds together here, in a quite different way, what rays out finely without arranging itself flower-like around a center.

If we also note the way in which yarrow does not use up all its resources in flowering, but continues to send out rosette-forming runners in all directions underground, and the way in which from them, roots, that remain threadlike, penetrate the subsoil, then all of this further extends our picture of its capacities which ultimately are connected with its seeds as 'possibilities'. (See pictures in the appendix on page 156ff.)

Summarizing the picture of yarrow

In summary, these capacities can be expressed vividly as follows: yarrow penetrates the subsoil, feeling its way sensitively downward, and radiates upwards vegetatively, becoming increasingly concentrated, and is stopped at the level of the flowers as a whole. Flowery astrality has become concentrated in the vegetative part in the expressive power of a limiting factor, and not as a radiant flower form as in an umbel.

The soul-spiritual environment that is reached this way gives the inner connection with the human being; the part of the plant that we harvest (inflore-

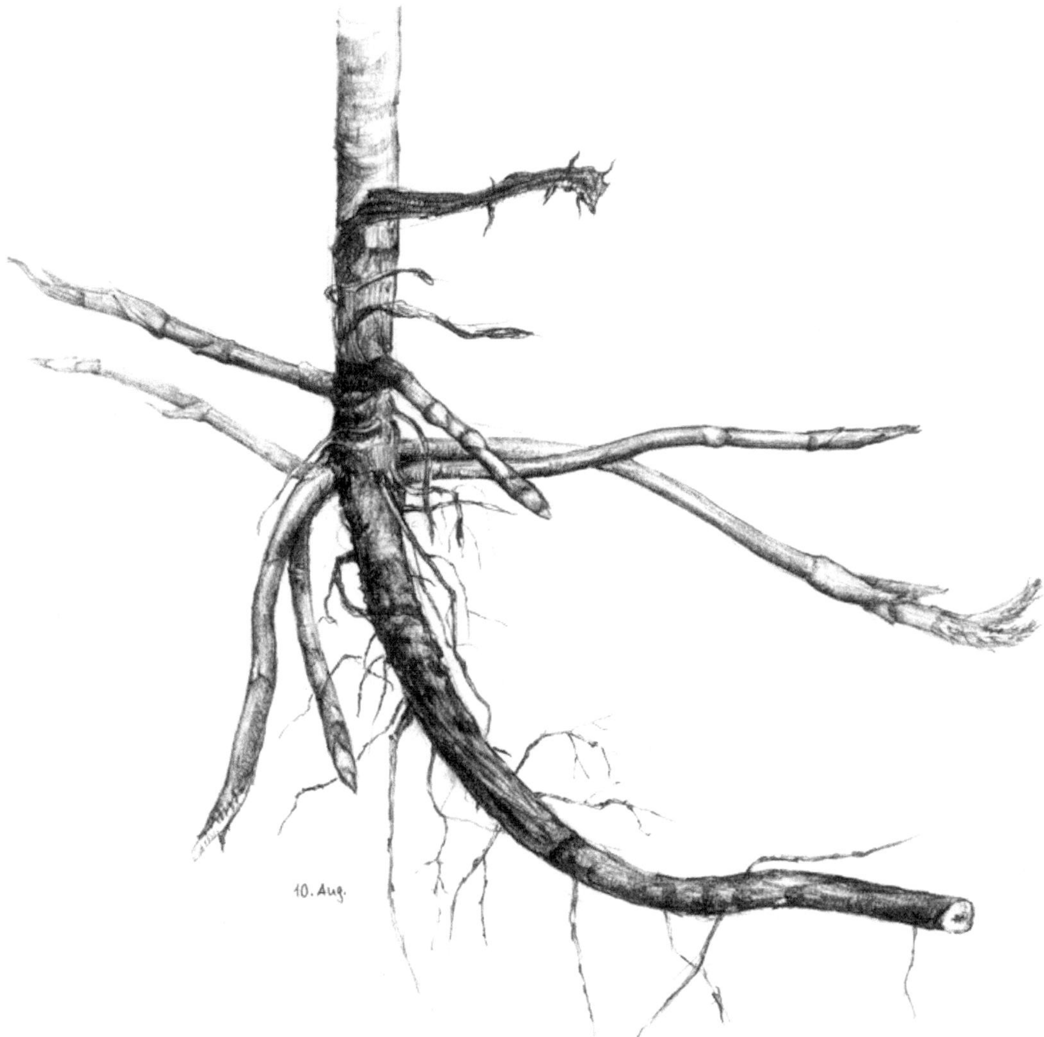

scence and upper leaves) can be taken as an image of the way that the continuing rhythmic life process of the human being consolidates towards the metabolic aspect. (See 'The Experience of the Threefold Organization', page 87) Depending on how they are used, these substances support capacities. They stimulate digestion where metabolism begins (stomach, intestines, gall bladder and pancreas); inhibit inflammations, promote haemostasis and stimulate renal processes via the fluid element (concentration during secretion).

Runners formation in yarrow at the transition between shoot and root

Chamomile leaf

Building up a Picture of Chamomile

After working with chamomile for some time, I contemplate its development:

1. I imagine the way in which its delicate rosette in spring tastes more like a sweet herb, like parsley;
2. the plant then passes over into growing profusely and acquires a somewhat sharp to bitter taste like horseradish, but still does not smell like chamomile;
3. it then loosens up in all directions, becomes transparent and now tastes somewhere between lettuce and fennel;
4. it finally flowers and opens up to its surroundings and begins to emit scent.

Here again these processes can be consolidated as we experience them by means of constructing a symbol based on observing a leaf and drawing it. I connect the taste change from sweetish like parsley to sharp and bitter and then like fennel-lettuce with the fine threadlike sweeps of its feathers that are loosely opened to the environment. Thus the fundamental gesture of a hidden inversion of chamomile becomes ever more clear to me in inner perception. This gesture becomes an outer picture at a higher level that marks the transition from sharp taste to aromatic odor through the way the flowers manifest doming and hollowing of the receptacle and the reflexing of the ray florets. Thus, the characteristic honeyed, pungent scent of chamomile arises only in both flowering and fruiting as a metamorphosis of the sharp and bitter taste that for a time was concentrated in the growing foliage. Finally, the plant spends itself completely in the scattering of its tiny seeds. Up till then only the tap root, which in its upper part thickens somewhat as in a turnip and from which secondary roots emerge, constitutes a principle that continues throughout.
As a bridge to the human being, the soul-spiritual environment of this chamomile gesture leads us

Chamomile flower

into a realm that is more separated from vital processes than in yarrow. Its anti-inflammatory effect is transmitted directly to mucous membranes by the moist heat of steam as a kind of taste. In tea form its fruity, honeyed taste stimulates the forces of the astral body and the I that overcome substances in the digestive process, so that they can eliminate cramps and pain.

It is only in the drying process that devitalizes the flowers that the scent of chamomile really comes out. But with yarrow, where the flower nature concentrates in the vegetative, I would prefer it fresh rather than dried. (See appendix: Smelling and tasting exercises on yarrow and chamomile.)

Other kinds of chamomile (e.g. see page 132) have similar leaves but lack the characteristic scent and the inversion is not as pronounced. It is only with true chamomile that I can rediscover the gesture, previously found in the course of development, in the very different sorts of experiences with smell and taste of the flower and fruit region. Thus, in such a study, it is not permissible to draw conclusions in one phenomenal region based on what goes on in another. It is only in the inner connections of what is studied that the power arises to arrive at what is specific to it. (See page 170 for aspects of chamomile root preparation.)

Rudolf Steiner on the use of chamomile flowers:
Now the interesting thing is that, so long as one has the nervous disease confined to the alimentary canal, or indeed, in the region of the stomach – that is, when it does not extend beyond the digestive organs – one can achieve the same result by injecting chamomile instead of arnica . . . chamomile contains practically no silicic acid, and you see from this that silicic acid first becomes necessary when one wants to act upon the nerves outside the alimentary canal. Silicic acid is not present, but sulfur is. This is essentially beneficial when it is a question of stimulating the etheric body in the digestive system.

Rudolf Steiner, *Physiologisch-Therapeutisches auf Grundlage der Geisteswissenschaft*, (GA 314), Lecture III, Zur Therapie und Hygiene, Dornach, 2nd January 1924, 2nd edition 1975, p. 215; trans: *Lectures to Medical Men* (London: Anthroposophical Medical Library). Typescript unrevised by lecturer

Rudolf Steiner on the use of chamomile roots:
The best remedial effects of these substances [sugar, alkaline salts and silicic acid – J. B.] will be obtained if you simply administer the roots of chamomile boiled in the appropriate way. It may surprise you that I speak of the root, but the different aspects under consideration here intersect, and we must realize that when the symptoms are severe, we need sulfur and what is taken from the flower organisation least of all.
What we really need is a substance that is still contained in a highly vitalized state in the plant, so that the long process it has to undergo will make the reaction vigorous enough. If we introduce into the digestive tract a suitable dosage of these substances as they are found in the root of the chamomile, the reaction [in the human organism – J. B.] in this case will not be strong enough to allow the vitalization to take place at the point of transition from the intestines to the blood; what is contained particularly in the sugar and silicic acid, but also in the alkaline salts, will simply be forced through in an untransformed state. Thus the kidney activity has a chance to absorb it into its radiations, and the substances absorbed in this way are then impelled by the kidney activity toward the nerve-sense activity, which is thereby relieved of the astral functions.
If we really have insight into these matters, if we realize that this way of proceeding therapeutically leads to the most healthy results, much can be discovered. Furthermore, we can very easily be led to other things. We can see how what is absorbed is transformed in the human organization, how the activity of the kidneys sets to work, receiving what is supplied to it by the channels of the blood and radiating it out; we can see how the plastic activity then reacts in its turn. Then we begin to see how this plastic activity in its pure form is restored by the administration of silicic acid, sugar, and alkaline salts.

To supersensible vision, silicic acid, alkaline salts, and sugar, mixed in the right proportions and viewed intuitively, form a kind of human phantom. Something

Comments on the uses of chamomile

With *chamomile*, we usually use the *flowers*, so that we are using plant substance that in its formation has reached the climax of differentiated organ development. This substance is no longer developing and it is best to administer it directly to the part of the human organism that is receptive to its effect, i.e. where the organism has to overcome the substance. However, Rudolf Steiner also gave an indication for a root preparation. It is more difficult to decide when this should be harvested. Our attempt to understand this is oriented on a general picture of root development. A plant root lives in the double stream of ascending and descending saps. In the course of its development it goes through various phases. These run parallel with phases of development above ground. Thus the orientation of the respective root processes also becomes observable.
One could describe these phases as follows:

1. *Embedding*: the main root pushes straight down into the depths and a rosette forms above ground.

2. *Spreading*: secondary roots and root hairs move out in all directions, usually in parallel with the more luxuriant growth of the vegetative organs.

3. *Fruit formation*: the root thickens like a turnip and accumulates substances that can be drawn on later (starch)

4. *Sprouting time*: the roots often penetrate deeper.

5. *Solidification*: the root lignifies and gives support to the plant above ground.

These phases can intersect in various ways. In *chamomile* after embedding (phase 1) phases 2, 3 and 4 occur almost simultaneously. Then the root lignifies (phase 5) as the plant begins to flower and to wither from below up.
Thus, as regards the root, Rudolf Steiner speaks of the presence of silica in it, of it being furthest

distant from sulfur and of what comes from the flower organization.
This is the exact opposite of the explanation for the use of flowers in the preceding quotation, where sulfur and the absence of silicic acid are emphasized. The extent to which silicon is important in chamomile can be gathered from Rudolf Steiner's agricultural course (1924). There we are told that plants like yarrow and chamomile, which have finely divided shapes and whose flowers have very little silicon by weight, nevertheless promote a receptivity for the 'finely divided silicon in the atmosphere' when they are used as compost preparations.
For the right effect of the chamomile *root preparation*, it is obviously a case of starting at the other end of flower formation, i.e. where the saps rise from the earth below into growing and blooming. Thus the substance is neither herb nor flower. It is *on the way* from the root to the transition of growth to blooming. What can be read afterwards from the way the plant is formed is present germinally, as it were, in this state of the root. An indication of this is already given by the scent and taste (see page 178). When the root strives towards flowering, at the stage of greatest activity of shooting and growing of the vegetative organs, and becomes somewhat tender and turnip-like, it starts to become fragrant and taste sweet. In our opinion this is the right time to harvest it. Both percepts disappear when the plant flowers.
A comparison of the smell and taste of flowers and roots at the right stage (phase 3 in connection with 2 and 4) can give us an impression of something finished and closed off (flower) in contrast with something that is conveying things (root).
If the conformation of the substance is retained during the manufacture of the preparation it can stimulate the human organism to complete in its own way the process that is broken off in the plant, i.e. to work in from the nerve-sense system instead of beginning directly in the metabolism. (See appendix for harvest time.)

like a phantom is there before us if we picture these substances in their formative force. They are preeminently sculptors, these substances; they bear the plastic principle within them. This is evident even in their outer formation through intuitive vision. (. . .) Those who have no sense for this artistry in nature – an artistry through which the organism is shaped, shaped chiefly by the nerve-sense activity – cannot discover in any rational sense the relationships between substances in the outer world and what is taking place in the human organism. Yet there is indeed a rational therapy – a therapy that is simply able to perceive processes that take place in the outer world, that are broken down in the human organism and can then be radiated out by the kidney activity and taken hold of by the plastic activity of the nerve-sense organism.

Rudolf Steiner, *Physiologisch-Therapeutisches auf Grundlage der Geisteswissenschaft* (GA 314), Lecture IV, Anthroposophische Grundlagen fur die Arzneikunst, Stuttgart, 28th October 1922, Dornach , 2nd edition, 1975, p. 152; trans. (edited): *Fundamentals of Anthroposophical Medicine* (New York: Mercury Press, 1986), pp. 75–7

Concluding Remarks

This volume of the guide ends with the examples of yarrow and chamomile and is published as working material.
With this first volume, a methodical approach is offered for developing a rationale for a knowledge of healing substances that is called for not only by Rudolf Steiner but also by the facts. As a result of this approach, new questions arose and these are pursued in Volume II and III.
Volume II (2000) contains a new understanding of chemistry and the human organs as a basis of a differentiated approach to healing illustrated by selected healing plants and compounds. This is followed by examples of pharmaceutical processes.
Volume III (2003) contains comparisons of approximately one hundred representatives of the families Umbelliferae, Cruciferae and Ranunculaceae, with colored pictures, leaf sequences, botanical details and traditional uses in herbal medicine. It also contains an index of all plants referred to in the three volumes.
Since we are dealing with working material here, again I would like readers who are trying to work with it to tell me about the experiences that they have had, and where problems or ideas and questions that supplement the work have arisen.

Appendix

Appendix

Material has been compiled here for those who would like to do more work with yarrow and chamomile. This material would overburden the main presentation in the guide, but they may supplement in various directions the picture that we can make of the two plants.

Appendix Contents

Supplementary Observations from Experiments at our Research Institute

Supplementary Observations from Experiments at our Research Institute

Yarrow

Comparison of the extreme forms of the Achillea group, A. macrophylla and A. ptarmica, with the middle form, A. millefolium, which underlies this study.

See Jochen Bockemühl and Kathleen Clark; 'Beiträge zum Verständnis der Schafgarbe', *Elemente der Naturwissenschaft*, **25** (2) 1976, 22–34

Development in the Course of the Year

Achillea millefolium: After germinating in the spring common yarrow forms a rosette. This develops increasingly large leaves that spread out on the ground in an open situation or, in a shady situation, remain upright, for instance in tall grass. The main stem shoots up in the middle of June and develops its first buds. At the same time side-shoots develop and, in sunny places before midsummer, also become flower shoots or remain sterile. The whole plant flowers mainly in July and August. Side-shoots can also produce inflorescences in September, October or even November. After the plant withers, the firmly erect, lignified stems remain standing for a long time.

In the sequence of leaves from seedling to flower on the main stem, the first, small leaves are still somewhat flat and barely divided. They soon change, becoming larger and finely divided as they approach the typical longish oval leaf form. The branched pinnations also begin to spread from the petiole. Then the leaves slowly become smaller and more pointed outwards. The emphasis of the 'blade surface' moves down to the leaf base. The last leaves near the flowers become narrow, linear, pointed and only slightly serrated.

The very characteristic somewhat bitter taste of yarrow is at its bitterest and somewhat 'leafy' in the lower leaves of a flowering shoot, and it becomes somewhat more aromatic in the topmost leaves. The leafy part of the taste predominates in leaves that are still growing or have grown in the shade. In the single leaf the top is usually more bitter than the base. The bitter-aromatic taste is especially concentrated in the flower heads.

The taste and smell tests described were done on freshly crushed parts of natural or cultivated yarrow, and the findings of several people were compared. Although the white to reddish yarrow flowers shine very distinctly out of the dark green, they have no strongly emanating scent, no eye-catching differentiated color, and they do not present a particularly individual 'face' as does a poppy or a daisy. Other flowers radiate something towards us that attracts our notice and draws our attention closer. For instance, we quickly notice elder flowers through their characteristic, fresh, pungent, overpowering scent, and we can hardly look at willow catkins without imagining ourselves stroking them. Compared with them, yarrow flowers remain much more neutral. On closer inspection what at first seems like a dense, flowering surface proves to consist of lots of little flower heads and these in turn of tiny individual flowers. Thus a high degree of differentiation and grouping together is attained in the Compositae. But although individual flower heads look like *one* flower through the placement of tubiform florets in the center and ligulate florets round the margin, the whole inflorescence does not acquire an overall shape as do, for instance, the Umbelliferae. In these, single flowers on long stalks ray out from a point so that we get the impression of a higher unity. But in yarrow inflorescence the branching progresses irregularly, and the grouping together only takes place at the flower level.

Unlike the leaves, which present an overall form in spite of the mobile, branched pinnation, the finely membered structure of the dense cap of flowers just stops somewhere at its periphery, without taking on

Common yarrow (*A. millefolium L.*)
Sowing: 7 November
Developmental stages: left: 14 June
right: 1 August

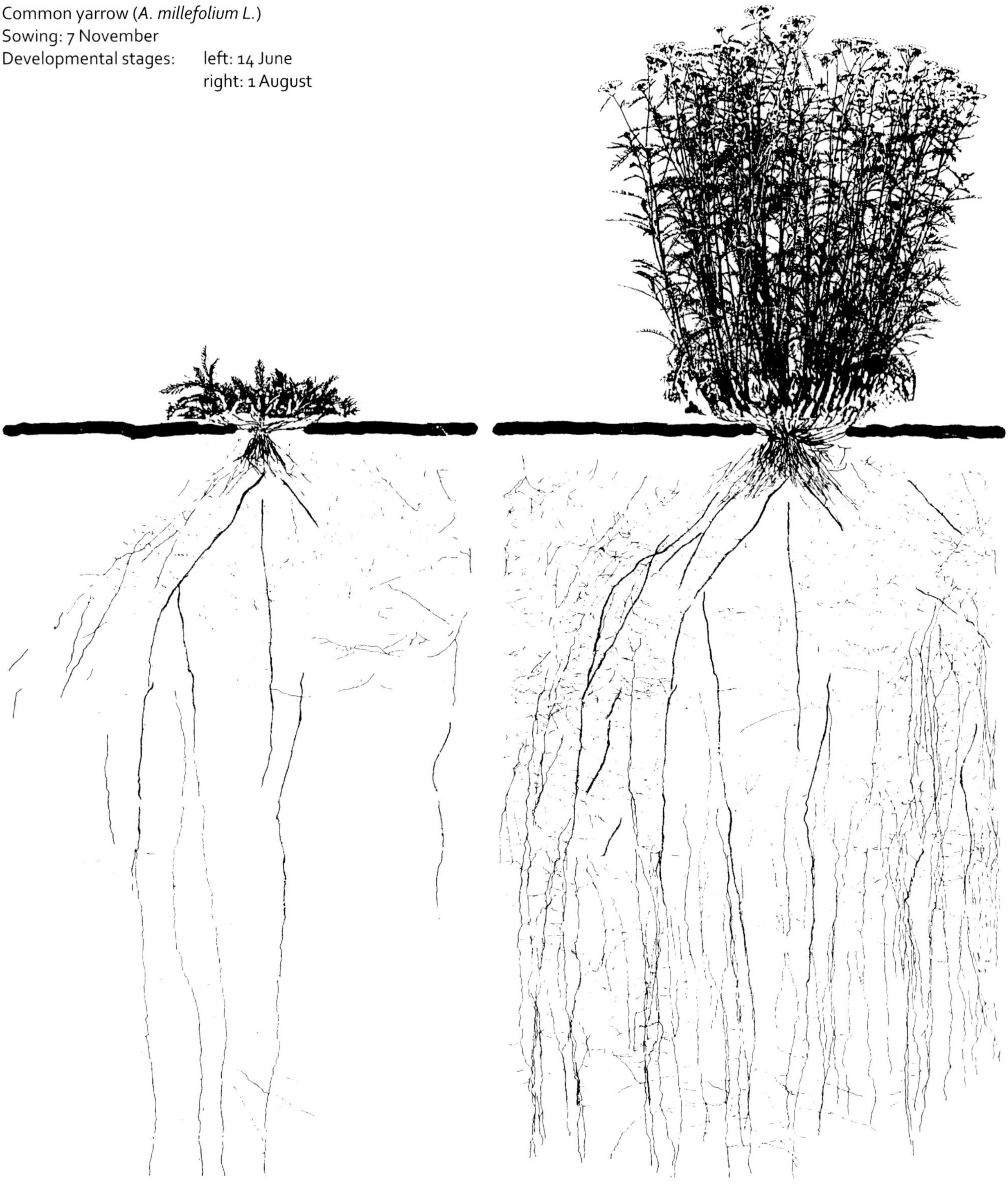

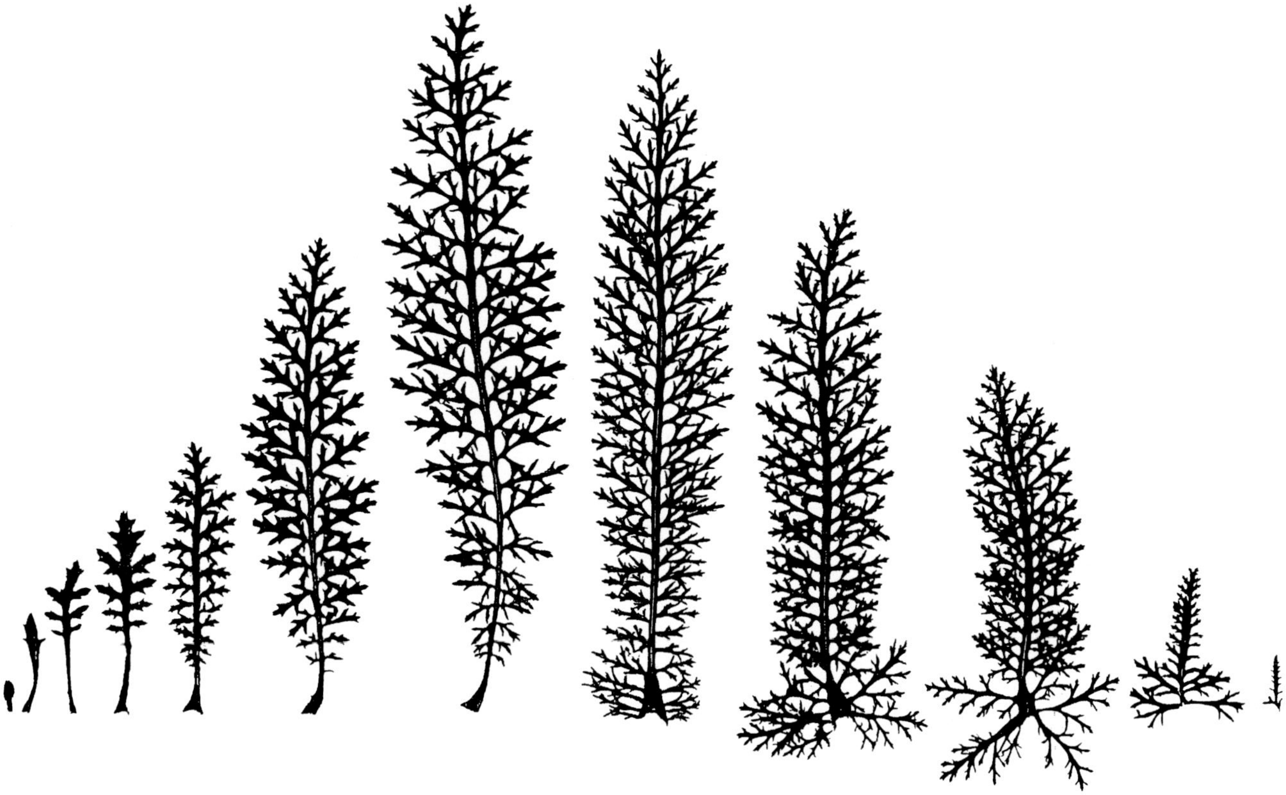

Achillea millefolium:
Selection of typical leaves from seedling to flower

any particular shape.
We will start with creating a general picture of the root region from which we can experience that a continuous, uniformly flowing growth movement weaves nimbly through the soil in manifold interwoven threads. An intimate connection with the soil and water in it arises. This makes the plant even more tightly connected with the earth – such that we cannot pull it out without damaging it.
Movement and firmness of this kind are especially pronounced in yarrow root. The roots that go out from the base of the plant remain thin and they spread out further down in wavy forms. With their multiplicity of branches they form a dense network penetrating the soil.
The many runners that grow to all sides, above all in autumn and winter and with which the plant spreads like a lawn, are characteristic of yarrow. In spring they grow up toward the sun again and send fine root threads down into the depths.

Sneezewort (*A. ptarmica L.*)
Sowing: 7 November
Developmental stages: left : 19 June
right: 12 August

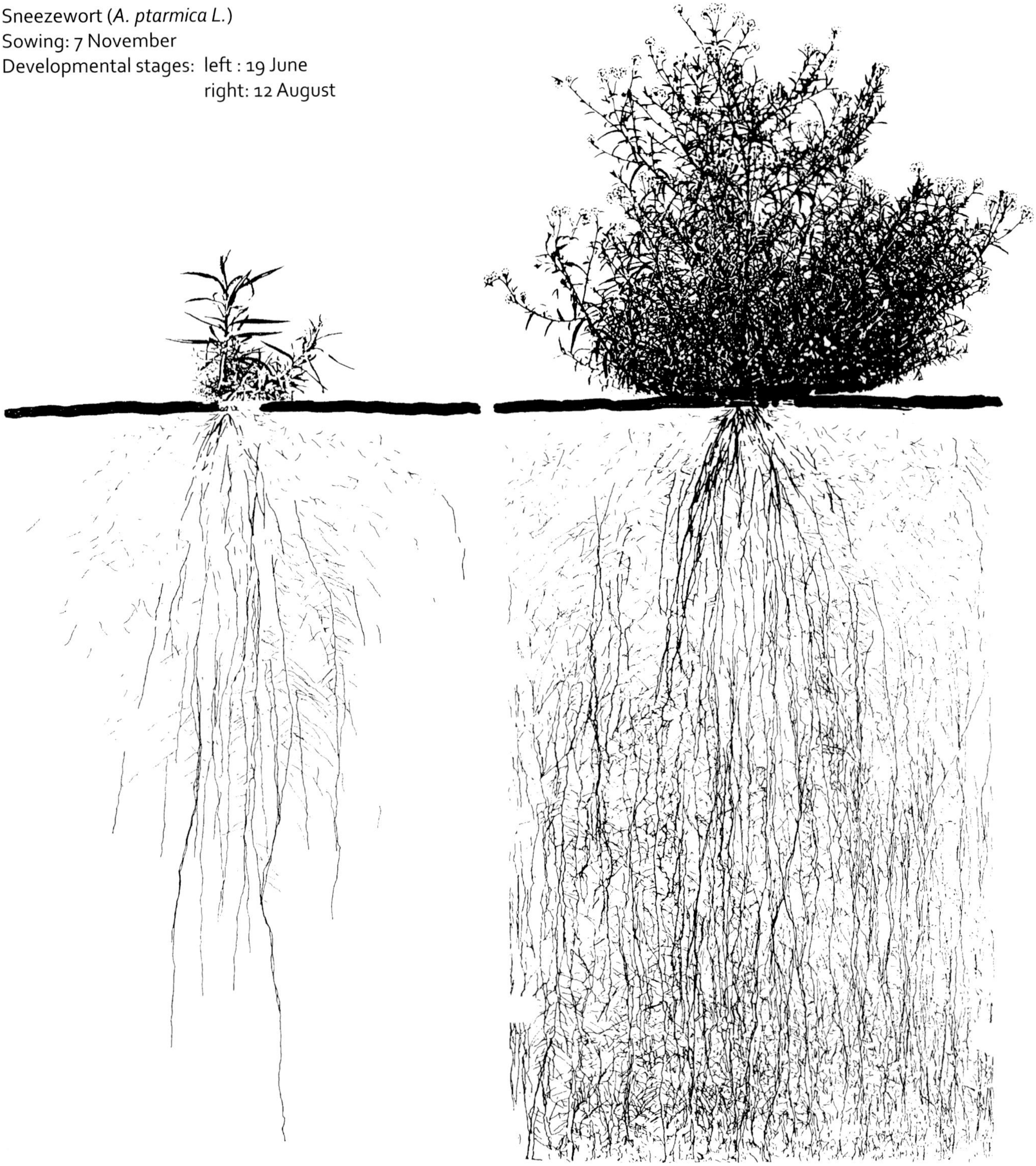

Achillea ptarmica L.
Selection of typical leaves from seedling to flowers over two vegetative periods

Sneezewort (*Achillea ptarmica*) forms practically no rosette. It generally shows a stronger tendency to flowering. After it germinates, its leaves point upwards and the shoot begins to rise very early. The main shoot shows its first buds early in July. The main flowering goes on from the beginning of August to mid-September. Like *A. millefolium*, *A. ptarmica* forms numerous side-shoots. In a root chamber, the plant became as luxuriant as a bush, with a very woody and brittle stem. The way that the whole plant rapidly stretches upward is again found in the gesture of the flowers. They do not stay on one level in an umbrella-like canopy but, with fewer though larger flower heads, form inflorescences like bunches of grapes. Like chamomile, they show the more centered characteristic of typical 'flowers'. But both the inner tubiform and the marginal ligulate flowers of the flower heads are white. The leaves of the leaf sequence of the main shoot are rather similar to each other. The first have petioles and the blades are rounded and scarcely separated from the petiole. In the sequence, the emphasis on blade expansion soon shifts down to the base of the leaves. On the whole the leaves all remain rather narrow like those of *A. millefolium* near the flower, but bigger.

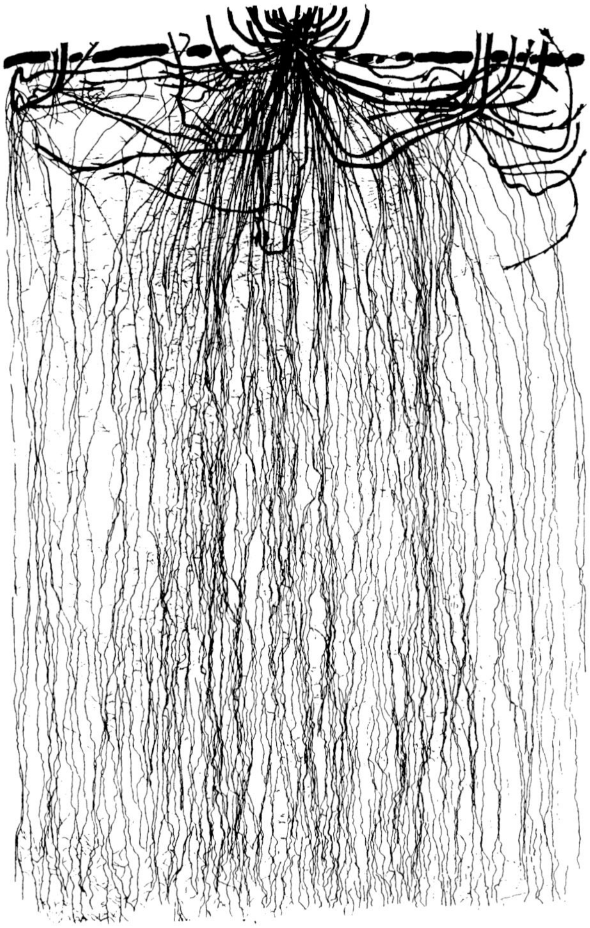

A. ptarmica roots and runners after a further vegetative period, 28 October

A. ptarmica leaves have no bitter smell or taste typical of yarrow, but the rhizomes that go out from the base of the plant (see adjacent picture) have a hot, peppery taste. The intensity of the taste decreases towards the top and this runs parallel with the 'shift' in leaf form of *A. ptarmica* versus *A. millefolium*. Its flower heads are not bitter-aromatic like those of *A. millefolium*. They emit a delicate, sweetly aromatic scent.

Its root grows straight down into the earth, just as its upper part grows directly up. Many finer roots grow straight down at the same time. The numerous, equally fine horizontal branches are much less obvious than the perpendicular growth. The roots – unlike those of *A. millefolium* – have a very sharp, strong taste. In autumn and winter, in contrast to the very fine root threads, thick, watery and very hot tasting runners form, which used to be dried to make snuff.

Broad-leaved yarrow (*Achillea macrophylla*) has broad, soft leaves with very pronounced and coarse pinnation. Unlike yarrow leaves previously described, here all the leaves have long, strong petioles. Thus after a contraction in the first winter, the leaves become more strongly divided and the pinnation extends to the bottom of the petiole.

The leaves slowly become smaller until they take on a coarsely serrated, linear shape near the flower. *A. macrophylla* grows more slowly than *A. millefolium* or *A. ptarmica*. Although it only began to bloom in the second year in our experiment, there were never more than four or five basal side-shoots. These shoots were relatively large, however. The flowering shoots grow at the beginning of May, develop firm buds and then bloom in umbrella-like inflorescence through the summer.

Leaves that grew in the sun had a strong scent reminiscent of yarrow and a strong taste. Plants grown in somewhat shady, naturally wet mountain spots were somewhat more delicate and during flowering had a weak scent and a mostly leafy taste. In the predominance of the roundish, flat, stemmed form and in the somewhat revolting, leafy yarrow-like taste we find something similar to the very first leaves of *A. millefolium*. The pinnation is very similar to the one we noted in *A. millefolium*, except that the flat principle of the lowest leaves is retained somewhat in the following leaves where it makes the blades wider.

A. macrophylla roots grow slowly, with a few thicker root strands that mainly grow straight down. There are fewer side branches. The small, thick, horn-like formations that come out of the rhizome correspond to the runners of the other two plants. They stay quite near the base of the main shoot and they show the same contracting tendency as the root.

Broad-leaved yarrow (*A. macrophylla L.*)
Sowing: 7 November
Developmental stage: 12 June

Broad-leaved yarrow (*A. macrophylla L.*)
Sowing: 7 November
Developmental stages: left: 2 August
right: 14 October of following year

Achillea macrophylla L.
Selection of typical leaves from the first and second year

Connections with the Environment

Although common yarrow prefers loamy soil and sunny spots in the temperate zone it also tolerates a variety of other conditions, providing that there is enough light and a little moisture in the soil (see Heinz Ellenberg, *Vegetation Mitteleuropas mit den Alpen*, 4th Edition, Stuttgart, 1986). Thus it is spread out over the whole Northern Hemisphere in mainly sunny places in numerous varieties and races that are hard to distinguish. It is also often hard to distinguish the *Achillea* species themselves, because a further play in the outer phenomenal picture is connected with the ability of the plant to adapt (see page 135).

This ability to adapt gives *A. millefolium* a central position among the species that are limited to a narrower territory. In addition to the species described here we also have in mind other *Achillea* species including the extremely small, white, downy mountain forms such as *Achillea nana*, and forms in the hot, dry Mediterranean climate. Plants of the same origin that were grown in sunny and shady (i.e. away from direct sunlight) positions show us the direction in which the changes in shape and development can go, and also the characteristic things that are retained. These observations are in good agreement with those made in the natural habitats.

In semi-shade they only form a few long, fairly loose, narrow pinnate leaves with long stems. They are very delicate and light green. The oval outline is well preserved here. By contrast, the leaves of plants grown in the sun are much smaller, more numerous and more concentrated. This is especially visible in their formation and firmness, and in their dark green color and the very dense pinnation that feels 'dry'. In the sunny position, numerous firm, stiffly upright basal flower shoots form. In the shade they remain soft, small and with hardly any flowers, or they are absent altogether. The relatively weak growing power is mainly concentrated in the rosette. The root growth of yarrow grown in the sun is also greater than of those in the shade. *A. ptarmica* prefers damp ground and lighted conditions, as its German name 'marsh yarrow' suggests. In natural conditions it mostly occurs on lime-poor riverside soils in very cold regions and on damp meadows and fields (Hans E. Hess, Elias Landolt, Rosmarie Hirzel, *Flora der Schweiz und angrenzender Gebiete*, 3 volumes, 2nd edition, Basel, 1996/1997/1980). This species is mostly found in the northern countries of Europe and West Asia (*Achillea*, Gustav Hegi, vol. 6, part 3, Hamburg, 1974).

Under the extreme conditions chosen here, *A. ptarmica* was more dependent on stronger sunlight than *A. millefolium*. It did best in the sun. In semi-shade it grew poorly or not at all (see figures on pages 166, 167).

The developmental tendency of *A. macrophylla* (shown on page 168) lay in the opposite direction. Here the vegetative plant grew very luxuriantly in semi-shade in the first year, although the number of leaves was somewhat smaller. The greenness of the leaves lightened in the sun. The second-year plants in both places flowered well. The preference of *A. macrophylla* for somewhat shadier growing places can also be seen clearly under natural conditions. It occurs on damp, fertile, humus soils at the edges of woods and in open scrub. It seems to avoid prolonged, direct exposure to the sun, but in the western and southern Alps, it clearly needs the intense light of the mountains.

▷

Form modifications of *Achillea millefolium* due to different growing conditions in a small area near Sils Maria (Graubünden, Switzerland, ca. 1800 m, July 1976).
A and B – flower shoots and basal leaves from a fertile meadow: A is an enclosed situation; B free- standing on disturbed ground. C and D – the same from an unfertilised meadow: C between sparse, tall grass; D surrounded by close-cropped grass. E – from a fertile position next to a cowpat near C and D. F, G and H – single leaves from non-flowering plants: F in shade, on moist, fertile soil; G in shade on somewhat dry, poor soil; H in sun on a rock with very little soil.

E
D
C
B
A
H
G
F
E

A. Millefolium

July, shade

July, light

May, shade

May, light

A. ptarmica

July, shade

July, light

May, shade

May, light

A. macrophylla

July, shade

July, light

May, shade

May, light

Sulfur and potassium in the plant process as a characterization of yarrow

In the following we will focus on two active principles that can be found in all plant processes. Their special kind of interaction will help us to understand yarrow. The principles will here be examined in connection with what Rudolf Steiner in his *Agriculture Course* (1924) called the way of working of *sulfur* and *potassium*. (*Geisteswissenschaftliche Grundlagen zum Gedeihen der Landwirtschaft* (GA 327), Dornach, 7th edition, 1984; trans: *Agriculture Course*: The Birth of the Biodynamic Method (London: Rudolf Steiner Press, 2004). He said that they are connected in such a special, exemplary way in yarrow that it can serve as one of the plants for the biodynamic compost preparations. We will try to formulate some ideas here that are intended as guides to the corresponding phenomena. (Other active principles such as carbon, oxygen, nitrogen, silicon, calcium, and their mutual connections, should be studied in the same way but we cannot go into them here.)

First: In the part of the plant process where matter is refined and given shape and becomes a picture of something spiritual that wishes to speak to us out of a center, an active principle is revealed which connects spiritual formative force with what is material, objective.
In plants, this is (generally) most clearly the case in the flower. But this principle, which here we should like to call *sulfur*, is involved in *every* formative process in a plant, where something ideal appears in sensory things. It comes to the surface only in the flower. There it leads to a releasing at the plant's outer limits, which can be crossed to varying degrees through the release of scent and pollen.

Second: We see the mode of action that Rudolf Steiner ascribes to *potassium* at places where the plant process is manifesting as swelling growth, so that bulk arises and is at the same time held erect and focussed on a concentration.
Rudolf Steiner points out that potassium holds back the growth for forming 'stem', 'rigidity', 'framework'. This active principle manifests at its purest in the vegetative growth of shoots and leaves, but it permeates the whole plant process. It is involved wherever plant substance is built up.

When we look for the active principle of sulfur in yarrow we are directed to the whole plant through the strongly radiating and yet neutral way in which it flowers. Here we find the matter refining principle that strives to appear vividly in the shaping of all organs, but especially in leaves with their fine, very mobile dividing which, in the overall shape, is nevertheless completely unified.
Here the sulfur process at all levels seems to be particularly connected with the active principle of potassium. This shows in a growth that does not flow and spread out but which aims at inner rigidity and concentration without the finely structured periphery becoming hardened. This makes the whole leaf look thicker, and the whole plant is somewhat rigid and 'dry' but not as crystallized out as thistle. Furthermore, we also find the sulfur and potassium principles connected where vigorous rosettes divide into numerous shoots that push stiffly upwards and lignify, and where mobile branching is stopped at a plane in the tight grouping of the flowers.
The tendency of runners to divide finely and spread out is also supported by the strong vitality. This indicates that the plant always holds back somewhat in the vegetative realm and never puts itself wholly into flowering. Under unfavorable conditions, for example in a meadow, yarrow can grow on for years and spread out like a lawn without flowering.
We can also find the gesture of yarrow in its smell and taste. Its characteristic smell is only released if we crush its leaves with our fingers. So we have to do something ourselves to get at the smell of yarrow. It does not release it willingly.

There is something contracting in the picture of the smell and nothing that streams out. The taste that is concentrated in the flower heads is somewhat bitter and also aromatic. This likewise points to a close connection between the two active principles. This special connection becomes even clearer by comparison with other *Achillea* species. Whereas in common yarrow the balance leans more towards the leafy region whose action radiates, so to speak, into the flowers, in which the aromatic, bitter taste is particularly concentrated once again, in sneezewort we find that the yarrow motif is pushed more into the flowers. It becomes a flowering shrub, in which the flower heads are bigger, begin to be fragrant and to break out of the closed confines of the planar inflorescence. The form principle of the leaves – likewise pushed more 'upwards' – seems to be more compact, pointed. In their contracted forms the leaves lose some of their finely indented outlines. The rhizome is more compact and the runners are thicker. The hot taste is concentrated further down. Sneezewort needs a lot of light and a moister soil. In contrast, in *A. macrophylla*, which likes shady, damp places, the yarrow motif is pushed down to the beginning of the leaves. This means that the spreading principle predominates in leaf formation. A luxuriantly leafy, softer plant arises. It does not form any real runners any more, because most of its vitality goes into the plant above ground. The leafy taste of specimens of these plants that grow in their natural, cool, damp, mountain habitat can, under warmer, cultivated conditions, give way to a bitter-aromatic, revolting smell that is easily released from its leaves. This makes *A. macrophylla* similar to the yellow flowered fern-leaf yarrow (*A. filipendulina*) that is often found in gardens (cf. parsley fern, *Cryptogramma crispa* L.). The shape and growth of their roots are also similar. Thus *A. millefolium* is seen to be a middle form between the two extremes. It connects the two together through the thematic similarity, but, in the differentiation that goes into every detail of shape and in the strikingly bitter-aromatic taste, it undergoes an intensification in which the interaction of the sulfur and potassium processes finds its most varied expression.

Chamomile

Chamomile in the Course of the Year

In 1970–72 experiments were made with monthly sowing in special chambers to facilitate root observation (see figures on pages 172 – 175). This allows the development of chamomile in the course of the year to be shown.

Chamomile usually develops its most luxuriant green when it grows upwards and works against flowering. It can do this all the more intensively the more time it has to form a long root while in its rosette form close to the ground. During the shooting, the rapidly growing leaves continue the tendency they already had at the beginning of development of dividing into fine, threadlike forms. They remain soft, wilt fairly quickly and make way for flowering. A picture arises of a life activity that takes place in airy space, which does not lead to hardening but metamorphoses into flowering and fruiting in accordance with the inversion described.

The root is active from seedling to flower formation. At first it mainly grows straight down. This tendency to go deeper is gradually accompanied by the growth of finer secondary roots.

At the time when the stem starts to shoot up, the more the root has been able to strengthen itself by penetrating deeper, the more intensively will the small secondary shoots grow. This is helped by the seasons. As the days shorten, development of rosettes and side-shoots as well as the downward growth of the main root predominates. The main root tends to store reserves during this time. In spring, as the days lengthen, the rosette grows a little more at first, then the root puts out more rootlets. Its vegetative, shooting development begins and passes over into flowering. Flowering and fruiting continue into autumn. If shooting and flowering begin in the autumn as a result of sowing too early (figure pages 172–173), the tendencies of the root to grow into the depths and of the organs to spread out on the surface of the ground intermingle. Then the flowers on the main shoot do not develop fully, and root formation is 'indecisive'. The root can strengthen greatly, but it prematurely loses its vitality again through the flowering. Only when chamomile can preserve some of this vitality through the winter does it begin to flower a second time in spring on the now upright side-shoots.

In contrast, plants sown in spring have too little time to strengthen themselves in the rosette, and so they remain weak (figure page 175).

Therefore, optimal development of chamomile occurs when the seasons unfold fully and properly in the corresponding developmental stages. We found that this generally happens for sowings from late summer to the beginning of autumn (end of August through September, see page 176).

Attempts to understand its root preparation and to determine the time for harvesting

It is easy to see when chamomile flowers have reached their peak, but with roots it is harder. Therefore seeds were sown at the optimal time (end of August) and roots were harvested at various stages of development. In 1969–71 and 1974–5, Carsten Schmock studied the overall shape and corresponding leaf sequence on the main shoots of chamomile together with the smell and taste of the root (unpublished student project).

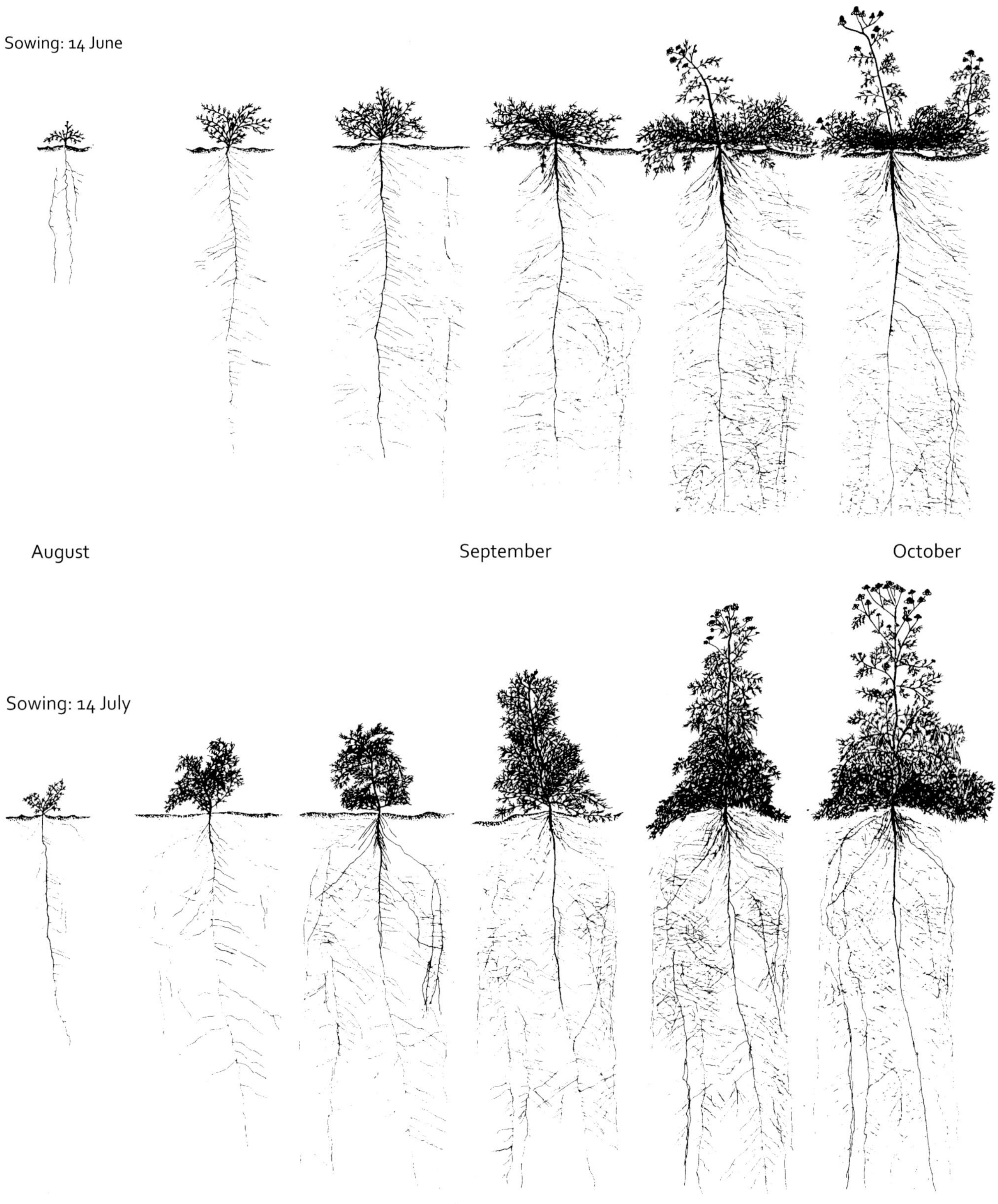
Sowing: 14 June
August
September
October
Sowing: 14 July

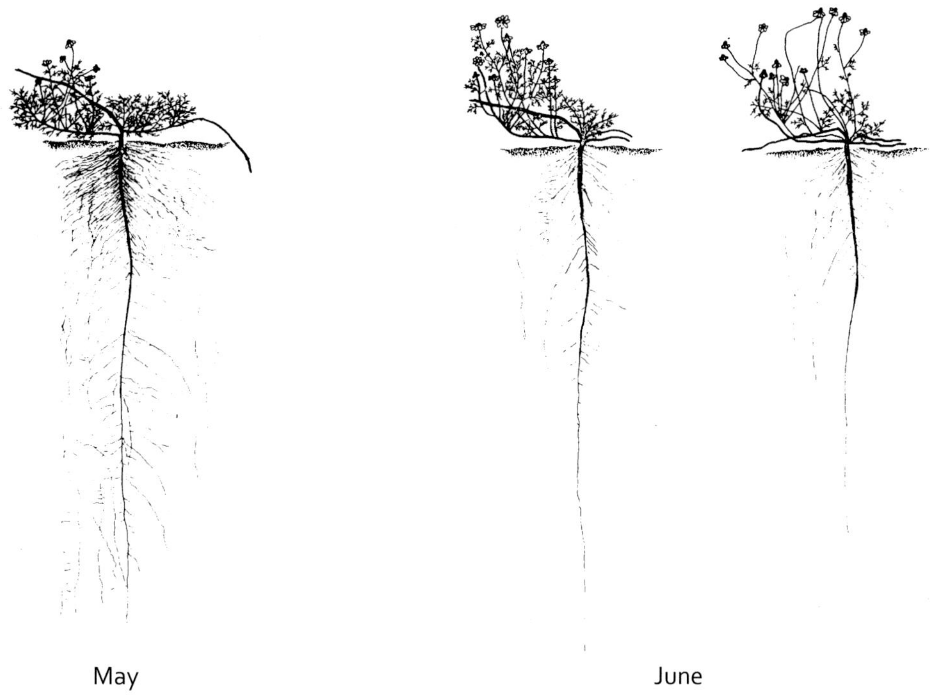

Chamomile: course of development when sown very early

For optimal development and spring sowings see following double page

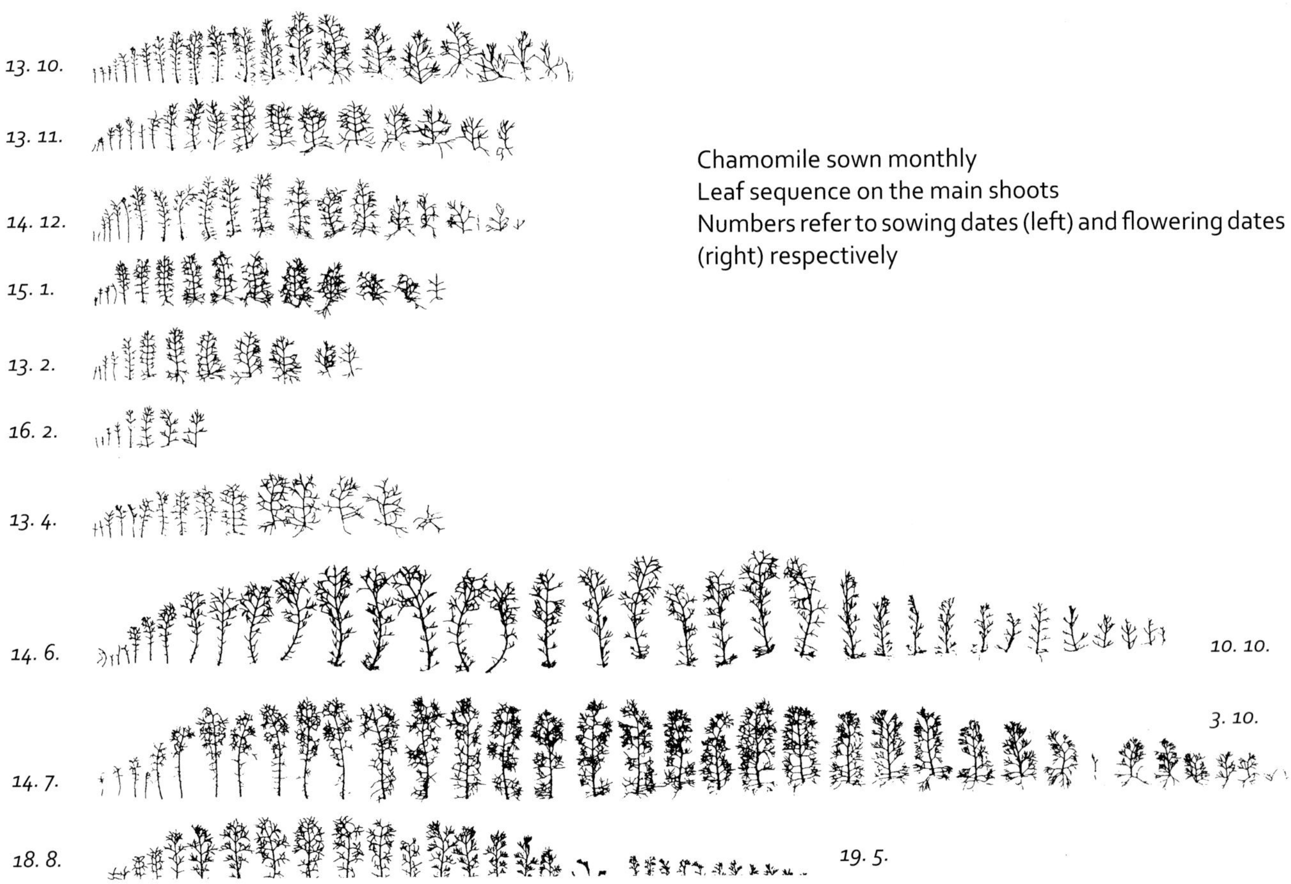

Chamomile sown monthly
Leaf sequence on the main shoots
Numbers refer to sowing dates (left) and flowering dates (right) respectively

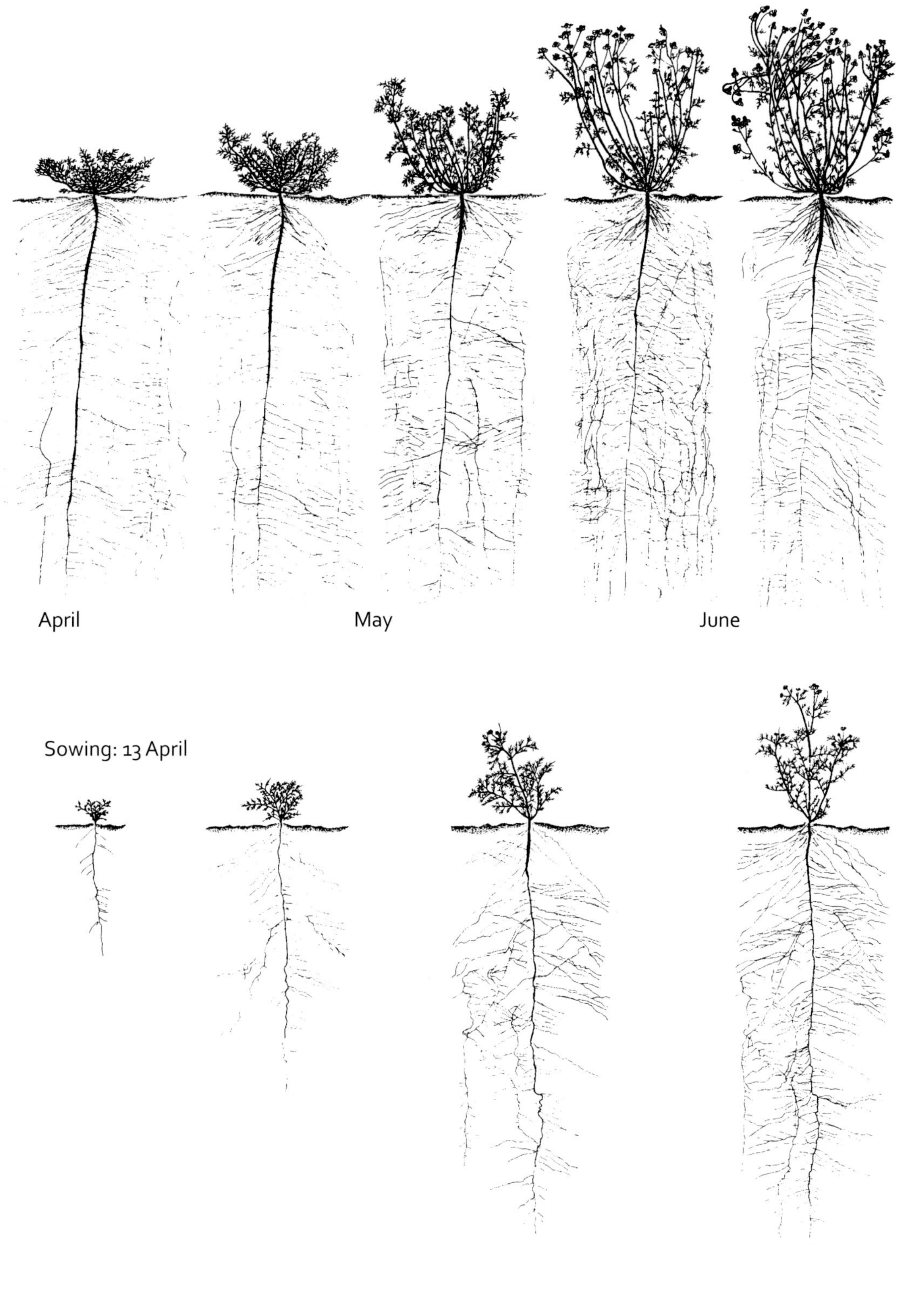
April
May
June
Sowing: 13 April

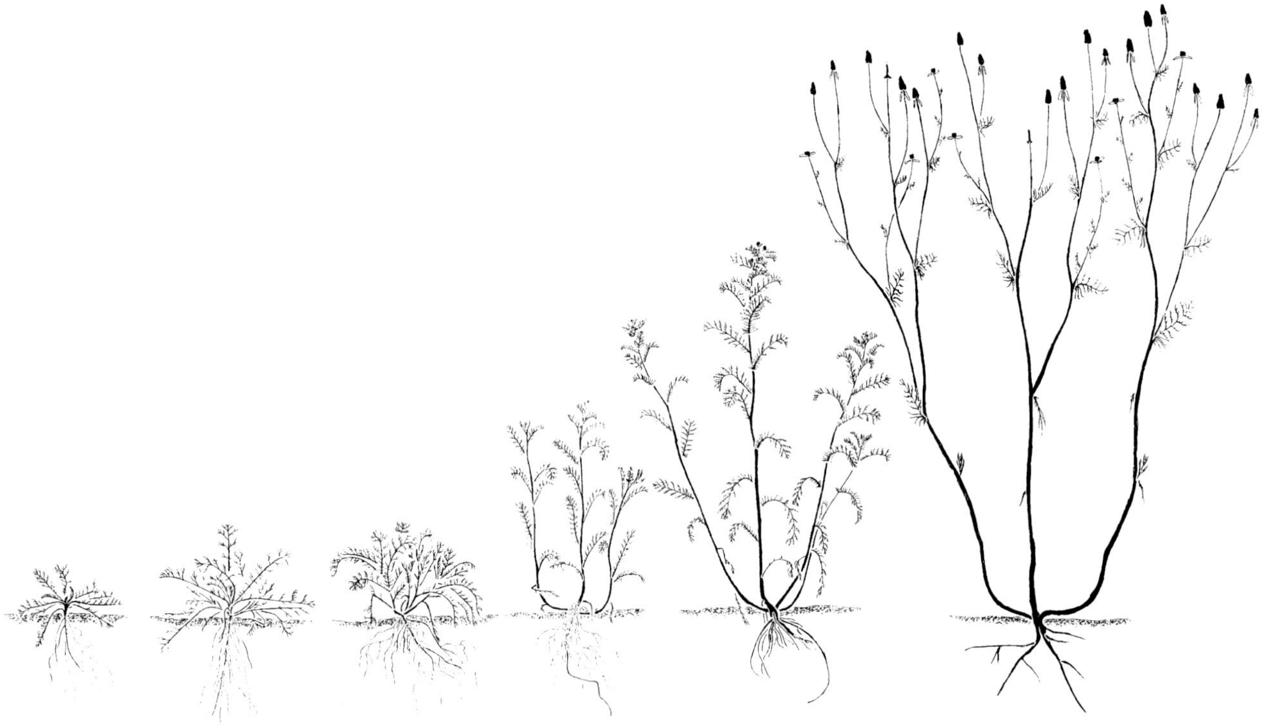

Developmental stages of the respective harvest times (roots not shown in full)

Sowing: 29 August

Stages recorded:

1	winter rosette	2 December
2	spring rosette	15 March
3	side-shoots present but not fully extended	2May
4	side-shoots beginning to lengthen	2 May
5	beginning of flowering	16 May
6	fading plant	8 July

The most vigorous root development was observed on 16 May at the transition between leafy proliferation and shooting to flowering.

			Stage 1 2.12.74	stage 2 15.3.75	Stage 3 2.5.75	Stage 4 2.5.75	stage 5 16.5.75	stage 6 8.7.75
Fresh plant	taste	leaves		like parsley	like horseradish young: like parsley		old: like lettuce green: like horseradish young: like parsley	sulphurous tannin-sourness
Herbal preparation	taste	leaves	like rooty hay		weakly aromatic		no taste	as 5 but weaker
							furry on the tongue	
Powder	smell	roots	somewhat bitter, leafy, pungent		neither somewhat bitter nor aromatic		sweet, aromatic, full	as 5 but weaker
	taste	roots	penetrating, somewhat bitter		full sweet		less sweet and aromatic	
Tea	taste	leaves	leafy aromatic open	like 1 but stronger	full leafy, slightly sweet			flat, fruity like fish oil
		roots	refreshing, stimulating		sweet, not fruity		fruity subdued less sweet	as 5 only weaker

The table shows that stages 4 and 5 are best for harvesting roots, because at these stages their smell and taste reach a high point. The best harvest time can be determined as much by developmental stage as by smell and taste. This is also the time when the root has freshly swollen like a turnip, but before it lignifies.

The roots were also analysed and crystallization pictures were made, but this will not be discussed here.

Smelling and tasting exercises on yarrow and chamomile (Ulrike Ortin)

Tea preparations (unknown origin) were each mixed with water in the ratio of one part of preparation to nine parts of water. With chamomile the flower heads were used, and with yarrow the flowers, upper stem and leaves.
The dry preparations were first hydrated by leaving them standing in two parts water for half an hour. This made the chamomile noticeably sticky and doughy.
After the rest of the water was added a macerate (steeping in cold liquid for half an hour) and a decoction (boiling for half an hour in a beaker covered with a watchglass) of each preparation was made.
The four preparations were then compared.

The *color* of the differently treated extracts of one plant species was about the same.

Chamomile: warm orange-ochre
Yarrow: faint reddish-brown

The following *smells* were noted:

Chamomile: cold: strongly aromatic (typical chamomile smell), sweetly concentrated (round, full, enveloping), 'light' – 'bright'
hot: slightly more bitter, earthier (almost unpleasant), honeyed sweetness, 'heavy' – 'serious'

Without knowing which plant parts and methods of preparation were used the first (cold) was identified as flower qualities and the second (hot) as leaf or root qualities.

Yarrow: cold: tangy, leafy (less aromatic, sweetish, scented and enveloping than chamomile – somewhat earthy, provoking attitude of avoidance)
hot: noticeably increased spiciness, sharpness is added. Here too in the comparison between cold and hot extraction there is a qualitative change from lighter to heavier and denser.

Unlike chamomile, whose flower quality came out particularly well through cold extraction, in yarrow the special spiciness and the sharpness in its flowers only asserts itself through hot extraction. So in yarrow its typical aroma can be brought out by this use of heat, whereas in chamomile heat threatens to make its scent disappear or to make it unpleasant.

The following *tastes* were noted:

Chamomile: cold: aromatic, full bodied (comparable to the smell)
hot: very bitter (no longer any scent quality)
Yarrow: cold; faint
hot: very bitter (far removed from scent qualities)

After yarrow and chamomile are heated, their tastes become quite similar. Only a strong bitterness remains.
One could ask how dependent the taste or scent quality of a watery plant extract is upon the cen–tration.
The taste does not leave us as free and it is not as pictorial as the smell (compare descriptions like 'cheerful' and 'serious'). A strong taste component can blot out all the others. This was particularly noti–ceable in the tasting of the underground runners of yarrow, which brought an attacking, hot, almost anaesthetising impression.

The plant teas were left standing in closed screw cap jars for a week. The scent development was as follows:

Chamomile:	cold: fermented, sour and typically aromatic scent hot: musty acidic, mouldy tendency
Yarrow:	cold: musty, unpleasant, pungent, sour – mouldy tendency hot: fermenting vigorously, *typically* scented

The different effects of the cold and hot extracts from chamomile flowers and yarrow tops continue right into the tendency to ferment and the intensification of the scent.
Chamomile's typical smell is strengthened by *cold* extraction, and yarrow's through the further deve–lopment of the *hot* extract.

Taxonomy and Constituents of Yarrow

Yarrow is a good example of how the forms of the *Achillea* group increasingly fragment and of how the boundaries between species disappear as we study them closely. The characteristics (and the active principles!) cannot be completely pinned down. We are always dealing with transitions. The contexts and the significance of details in the whole are only accessible to inner perception through comparisons by mobile thinking. Single discoveries (including those of so called active principles) are increasingly put in context against this background.
We can see how complicated the relationships are considered to be today with the following excerpts from Hermann Hager, *Hager Handbuch für pharmazeutische Chemie* (Berlin: Springer-Verlag, 1994), pp. 45–7; (the footnotes in the excerpt are omitted – J.B.):

'The genus *Achillea* L. comprises about 120 species (compare this with the description below of the *A. millefolium* group – J.B.) and it is divided into five sections:

- Millefolium section (ADANS.) W. D. J. KOCH, e.g. with *A. asiatica* SERG., *A. crithmifia* WALDST. ex KIT., *A. lanulosa* NUTT., *A. nobilis* L., *A. tomentosa* L. and *A. virescens* (FENZL) HEIMERL and *A. millefolium* group.
- Ptarmica section (MILL.) W. D. J. KOCH with *A. atrata* L., *A. clavenae* L., *A. erba-ratta* ALL., *A. macrophylla* L., *A. moschata* WULF., *A. nana* L., *A. oxyloba* (DC.) SCHULZ-BIP., *A. ptarmica* L., and *A. salicifolia* BESS.
- Achillea section, with *A. santolina* L., *A. santoli–noides* LAG.
- Only a few species belong to the Babounya section (DC.) O. HOFFM. (*A. fragrantissima* (FORSK. SCHULZ-BIP.) and the Arthrolepsis section (BOISS.) BOISS. An. (. . .)

Distribution: Eastern Mediterranean region and other parts of Europe and Asia Minor. Few species in the temperate parts of Europe, Asia, Japan and North America. Only the Ptarmica and Millefolium sections are represented in Central Europe. The *Achillea* section has a Mediterranean-Asiatic distribution.

Constituents: Essential oils, some with azulenes; sesquiterpene lactones especially guaianolides, azulenogenic compounds, germacranolides and eudesmanolides; amides and polyenes; flavones, flavonols and their glycosides, according to species C-glycosyl flavones, 6-substituted flavonol-3-o-glycosides and/or flavone-7-o-glycosides.
The occurrence of azulene formers (proazulenes) is not a general feature amongst the genus *Achillea* and is thought to be connected with the chromo–some number. Some individual species of the genus *Achillea* have variable chromosome numbers and can produce azulenogenic or non-azulenogenic plants. Proazulenes occur mainly in tetroploid species (*A. collina* and *A. lanulosa*) and in the diploid small

species *A. asplenifolia* and *A. roseo-alba* but they are absent in *A. moschata*, *A. ptarmica* and other diploid species and in octoploid *A. pannonica*. In the taxa there is great variability too. For instance the undisputed hexaphoid *A. millefolium* is free of proazulenes in some places but not all.'

Achillea millefolium Group

Taxonomy: The following species and subspecies are assigned to the *A. millefolium* group: *A. asplenifolia* VENT., *A. collina* BECKER ex REICHENB., *A. distans* WALDST. et KIT. ex WILLD. *ssp. distans*, *A. distans ssp. stricta* (SCHLEICHER ex. GREMLI) JANCHEN, *A. millefolium* L., *ssp. millefolium*, *A. millefolium L., ssp. sudetica (OPIZ) WEISS*, *A. pannonica* SCHEELE, *A. roseo-alba* EHREND., *A. setacea*, and other local varieties.
The taxonomic evaluation of the varieties that belong to this polyploid complex is carried out in various ways. It extends from the extreme view that all varieties are only subspecies of *A. millefolium* to the view that all varieties are independent species. The great difficulty in identifying individual plants tends to favor the all varieties view. The diploid or tetraploid varieties *A. asplenifolia*, *A. rosea-alba*, *A. setacea* and the hexaploid taxon *A. distans* are relatively easy to distinguish. But the tetraploid variety *A. collina* is often difficult to distinguish from hexploid *A. millefolium* and octaploid *A. pannonica*. The older literature only assigns the following subspecies to *A. millefolium* L.: *A. millefolium ssp. collina* (BECKER) ex REICHENB., *A. millefolium ssp. millefolium var. vulgaris* L. HAYEK, *A. millefolium ssp. pannonica* (SCHEELE) HAYEK and *A. millefolium ssp. sudetica* (OPIZ) WEISS.

Mistakes: Individual members of the *A. millefolium* group are with few exceptions morphologically similar and some of them hybridize easily, which is why a clear identification can only be made on typical samples. Identification is also made difficult by influences that modify characteristics, so that allowance must be made for mistakes. The identification of some species (*A. crithmifolia* WALDST. et KIT., *A. nobilis*, *A. odorata*, and *A. virescens*) which occur above all in south-eastern Europe is only possible on the basis of a detailed description of the species. Individual representatives of the Ptarmica section (e.g. *A. moschata*, *A. nana*) are also very similar to

the *A. millefolium* group, except that they have more than six ligulate flowers per head.
Constituents: The above ground organs of the plants contain essential oil that varies considerably in concentration in the course of ontogenesis. Rising during flower development it reaches a maximum of 0.8–1.4% just before the flowers open, and then it falls to 0.3%. The concentration of azulenogenic ingredients also increases during the vegetative period. Proazulenes occur in the glandular hairs of the composite flower but not in the secretory pores. Most of the essential oil occurs in the inflorescences, less in leaves and still less in stems, which agrees with the frequency distribution of the glandular hair on the plant. The roots contain no essential oil. Specific changes in the glandular hairs, characterized by the formation of giant glands and substantial increases in the number of those glands, point to the use of herbicides. Many herbicides also influence glandular cell metabolism. For instance, they can decrease or completely block the production of proazulene and the activity of dehydrogenases in the glands.
The roots contain up to about 0.003% polyenes and unsaturated amides such as pontica epoxide isobutyl amides of decadienic acid, decatrienic acid, decatetraene acid, cyclic amides such as 2E, 4E, 6Z-decatrienic acid-(2, 3-dehydro piperidide). 2, 4, 6, 8-decatetraenic acid-(2, 3-dehydro piperidide), and N-Isobutyl-2E, 4E, 8Z-decatrienic acid amide.
Minerals and heavy metals such as Si, Fe, Mn, Zn, Pb, Cu, Cr, Ni, Co, Mo and Be are accumulated in *A. millefolium* (0.001–0.3%). There is a correlation between the accumulation of elements in the plant and the mineral content of the ground and air. *A. millefolium* can be used to monitor the concentra–tion of copper and lead in the air.
Distribution: The individual species of the *A. millefolium* group are established in different areas.
A. asplenifolia: on damp meadows in Eastern Europe; *A. collina* on sandy and sunny steppe in Eastern Europe and on rough ground; *A. distans* on mountains, meadows with long grass, and in open woods and scrub in south, south-east and Central Europe; *A. millefolium* on pastures and fertile meadows in Europe; *A. pannonica* in dry grassland in South, South-East and Central Europe; *A. rosea-alba* in meadows in the southern margins of the Alps from Switzerland to Yugoslavia and in small areas north of the Alps; *A. setacea*: in dry grassland in Eastern Europe, the western boundary runs through Central Germany and Switzerland; fairly uncommon.'
At the level of active substances, yarrow is very similar to chamomile with respect to azulene formation. The species that is today universally defined as *A. millefolium* seldom contains proazulenes that are considered important, active substances. What this means for the overall composition is still an open question (also, for example, in regard to the therapeutic use of yarrow or chamomile). They would at most modify the character of the taste and the effect of 'yarrow herb' in the sense described in this guide. *A. collina* that is often used because of its proazulene content corresponds to the picture we sketched of yarrow.
Breeders use self-pollination and occasionally tissue culture to attain uniformity.

Taxonomy and Constituents of Chamomile

Many of the questions regarding scientific nomenclature in modern taxonomy and the determination of the constituents in chamomile (see the extract below from the *Hager Handbuch*) are interesting for the choice of plants and manufacture of pharmaceuticals, but they do not fundamentally influence the approach described here.

Excerpt from Hager *Handbuch für pharmazeutische Chemie* (1994), pp. 817–19; (the footnotes in the excerpt are omitted – J.B.).
'*Chamomilla recutita* L. RAUSCHERT

Constituents: The parts of the plant both above and

below ground contain essential oil, whose composition is determined by its belonging to one of the known chemodemes. According to the dominant bisaboloid in each case, commercial chamomiles are divided into the following types: bisabolol, bisabololoxide A, bisabololoxide B, and bisaboloneoxide. The oil contents vary during the life of the plant. An increase is usually observed during flower development. The maximum content of 0.3–1.5% is reached before full flowering. The oil content drops considerably when the flowers fade. The content of matricin and bisaboloids varies similarly, and so does that of *trans*-β-farnesene. The roots contain only small amounts of essential oil and no therapeutically relevant chemical constituents such as chamazulene and bisabolol. Root oil contains *trans*-α-farnesene, the isomeric chamomilla ester and en-in-dicyclo ether, chamomillol, b- and cis-caryophyllene, and caryophyllene epoxide.

The presence of the irregular monoterpene artemisiaketone in essential oil of chamomile flowers is of chemotaxonomic interest because of the problems with classification within this genus.

C. recutita is a salt-resistant species that appears to accumulate halogens. Bromide contents of >50 mg/kg are frequently encountered in chamomile flowers. This led to an increase in the permissable limit for bromide content to 150 ppm.

Distribution: As varying control of bisaboloids and matricin is genetically fixed and this chemical differentiation is correlated with geographical separation of this species, chemodemes occur in *C. recutita*. The distribution into chemical races enables the unequivocal classification of all known chamomiles according to their profile of active substances.
Bisabolol races: endemic in Catalonia/Spain;
bisabolol oxide A-race: Egypt, Central Europe;
bisabolol oxide B-race: South America;
bisabolone oxide: south-east Europe, Turkey.
In contrast the known occurrences of chamomile differ only quantitatively as regards all other chamomile constituent substances.

Breeding has given rise to chemovars that can be grown outside the original areas of natural occurrence of species concerned.'

I find it interesting that modern breeding of chamomile appears to correspond precisely with the tendency that in our experiment was taken by the forms which were produced through sowing too early (see appendix page 172). As an inherited characteristic a mixture of processes is formed that can be observed in a similar way in vegetable cultivars in comparison with their corresponding wild relatives (see Jochen Bockemühl: 'Vergleiche zwischen Wild- und Kulturformen zum Verständnis der Nahrungspflanze und zum Finden einer Zielrichtung für Züchtung', *Elemente der Naturwissenschaft* **39** (2) 1983, 1–14).

Some observations on the problems with modern taxonomy

In relation to the variability of natural forms and those newly obtained through breeding I will comment only briefly that the concepts of taxonomy (for example the concept of species) that are used to fix and define particular forms, on the one hand contradict the very idea of evolution (i.e. change) and, on the other hand, nevertheless represent a precondition for a proper understanding of evolution. Of course, the living world contains both forms, the seemingly always the same and the modified.

If, as in Darwinism, we ascribe change to chance then to be consistent we should never speak in terms of 'yarrow' or 'chamomile'. What would arise is entirely open. People would have to renounce the entire field of reflecting on thinking in the context of personal experience, an activity that is certainly assumed in modern Darwinism even if unconsciously. But this renunciation would make any kind of knowledge impossible.

However, this examination of modern science has a positive outcome. And that is, on the one hand, in physical (sensorial) reality we are always dealing with the here and now and ultimately can grasp only what accords with thinking and from it judge the changes, on the other hand, what we approach with the concept of the type as an inner perception, initially with the help of concrete forms, is connected with the sensorial facts in a far more mobile way than we previously realized, i.e. in the sense of pure thinking. Here we come to a crucial change in knowledge that is necessary, for example, for the implications of the application of genetic engineering in the breeding of healing plants.

The figures were compiled by Mathias Buess, partly from publications from the work of the Research Institute. Silvia Briner produced the figures in the Appendix. The authors of the other figures are: Jochen Bockemühl pages 58, 74, 78, 79, 80, 81, 89, 96, 97, 98, 99, 131, 133, 137, 143, 148 & 152; Sandra Beresford page 63; Silvia Briner page 134; Bente Pünther pages 135 & 136; Sophia Watschnadse page 57.